2010 Annual Meeting of the American Electrophoresis Society

(AES)

Topical Conference at the 2010 AIChE Annual Meeting

Salt Lake City, Utah, USA
7-12 November 2010

ISBN: 978-1-61782-135-6

Printed from e-media with permission by:

Curran Associates, Inc.
57 Morehouse Lane
Red Hook, NY 12571

Some format issues inherent in the e-media version may also appear in this print version.

Printed by Curran Associates, Inc. (2011)

For permission requests, please contact AIChE
at the address below.

AIChE
3 Park Avenue
New York, NY 10016-5991

Phone: (203) 702-7660
Fax: (203) 775-5177

www.aiche.org

Additional copies of this publication are available from:

Curran Associates, Inc.
57 Morehouse Lane
Red Hook, NY 12571 USA
Phone: 845-758-0400
Fax: 845-758-2634
Email: curran@proceedings.com
Web: www.proceedings.com

TABLE OF CONTENTS

(5a) DNA Electrophoresis in Inverse Opals

Nabil Laachi, Department of Chemical Engineering and Materials Science, University of Minnesota, Minneapolis, MN and **Kevin D. Dorfman**, Department of Chemical Engineering and Materials Science, University of Minnesota - Twin Cities, Minneapolis, MN

We investigate theoretically the electrophoretic motion of a long DNA molecule inside an inverse opal, a class of photonic crystals that consists of a face-centered cubic array of spherical cavities, connected through narrow pores. Our model is based on Zimm's lakes-straits (LS) model whereby the DNA chain is distributed over a fluctuating number of cavities, or lakes, and moves by transferring mass through the narrow pores, or straits. In the present case, the LS model is modified to account for the spherical confinement inside the cavities and the discrete set of allowable exit points from a given cavity. Our results agree with biased reptation theory in the limiting cases of low and high fields. The model also predicts a new regime at the transition between the low and high field reptation regimes that is not observed in gel electrophoresis. In addition to faster separation rates, the range of molecular weights in this regime extends well beyond values of gel electrophoresis. The gain in performance results from the well-defined confinement and limited connectivity of the spherical cavities in an inverse opal, contrasting with the weak confinement and high porosity of an agarose gel.

See more of this Session: Advances in Electrokinetics and Electrophoresis: Fundamentals
See more of this Group/Topical: Topical 3: 2010 Annual Meeting of the American Electrophoresis Society (AES)

Assessment of Microalgae Viability Employing Insulator-Based Dielectrophoresis

Roberto C. Gallo-Villanueva,[1] *Nadia M. Jesús-Pérez,*[1] *José I. Martínez-López,*[1] *Adriana Pacheco-Moscoa*[1] *and Blanca H. Lapizco-Encinas*[2,*] *Email: blapizco@cinvestav.mx*

[1] *BioMEMS Reseach Chair, Tecnológico de Monterrey, Monterrey NL, Mexico.*
[2] *Microscale Bioseparations Laboratory, CINVESTAV-Monterrey, Apodaca NL, Mexico.*

Abstract

Insulator-based dielectrophoresis (iDEP) is a microscale technique where particles are driven by polarization effects under a non-uniform electrical field created by the inclusion of insulating structures between two electrodes. In the present work, differences in dielectric properties (cell's membrane conductivity) of viable and non-viable microalgae, *Selenastrum capricornutum*, were employed to concentrate and separate a mixture of live and dead microalgae cells. These results demonstrate the potential of iDEP as a technique for rapid assessment of microalgae viability, unveiling the potential of this powerful technique for environmental applications on lab-on-a-chip platforms.

Introduction

In recent years, numerous fields including environmental have increased their interest in microscale bioseparation systems. This is due to advantages brought by miniaturization such as requirement of small sample volumes, faster analysis, reduced costs, greater sensitivity and portability. Insulator-based dielectrophoresis (iDEP) is one of such systems where particles are driven by polarization effects in the presence of non-uniform electrical fields created by the inclusion of insulators between two electrodes. It has advantages over the common DEP approach, which is with the use of microelectrode arrays, since there is no degradation of the microelectrodes and are less expensive to manufacture.[1]

DEP force depends highly on particle size and conductivity, in comparison with that of the suspending medium, and the gradient of the electric field applied. This means that two particles having the same particle size will have different dielectrophoretic response in the presence of the same electric field if they present differences in conductivity. It has been demonstrated that the cell's membrane state has an effect on the conductivity of cells[1] and therefore DEP can be used to manipulate viable and non-viable listeria,[2] yeast cells[3] and bacteria.[1]

In the environmental field, assessment of microalgae viability is of great significance due to the ubiquity of these organisms in nature and multiple industrial applications. For example, microalgae are associated with toxins in water blooms, ecotoxicological tests, and wastewater treatment.[4] Therefore, to determine viable cells of microalgae after an algaecide application or as an indicator of pollutant remediation has important implications. Nowadays, microalgae are receiving great attention because of their significant potential as a source of energy and CO_2 mitigation[5]. However, to explore, develop and implement these systems at the laboratory and field-scale depends in certain degree in methods that could promptly discern cell viability loss of the system. A rapid and reliable method to determine cell viability could improve and ascertain the success of all these biotechnologies. Considering the need for a technique or method that would allow for fast evaluation of the physiological/viability state of algal cultures, this study presents the application of iDEP to fractionate and concentrate viable and non-viable *Selenastrum capricornutum* cells, a unicellular eukaryote alga commonly used in ecotoxicological tests[4]. The work presented here is based on differences on membrane properties and have the advantage that dyes are no longer required once the technique has been tested and adapted. For the present work, a glass microchannel with cylindrical insulating posts was employed to dielectrophoretically immobilize and concentrate microalgae cells employing direct current (DC) electric fields. Experiments were carried out using live and dead microalgae cells, by varying the electric field in order to observe differences in their dielectrophoretic responses. Results showed that live and dead microalgae cells exhibit negative dielectrophoretic behavior when exposed to DC electric fields, where live cells exhibited a stronger response than dead cells.[6] Successful immobilization and concentration of microalgae showing viability differences was achieved, demonstrating the great potential of iDEP as a technique for rapid microalgae viability assessment.

Theory of dielectrophoresis

Dielectrophoresis is an electrokinetic transport mechanism driven by polarization effects on particles when subjected to non-uniform electric fields. In insulator-based dielectrophoresis, such field is produced when insulating structures are placed between two electrodes creating regions of higher and lower electric field intensity. The dielectrophoretic force exerted on a spherical particle is described by:

$$F_{DEP} = 2\pi\varepsilon_m r_p^3 \operatorname{Re}(f_{CM})\, \nabla E^2 \tag{1}$$

where ε_m is the permittivity of the suspending medium, r_p is the particle radius, E is the electric field, and $\operatorname{Re}(f_{CM})$ is the real part of the Clausius-Mossotti (CM) factor which accounts for polarization differences between the studied particle and the suspending medium. When low frequency electric fields are used, as in the case of direct current, the CM factor can be approximated to the real conductivities as described next:

$$f_{CM} = \left[\frac{\sigma_p - \sigma_m}{\sigma_p + 2\sigma_m}\right] \tag{2}$$

where σ_p and σ_m are the particle and suspending medium conductivities respectively. From equation (2) it can be inferred that two particles with different dielectric properties will exhibit different dielectrophoretic response under the same circumstances of suspending medium and electric field applied. Moreover it can also be deduced that there are two types of DEP; when the particle has higher polarizability than the medium, the CM factor will be positive and the particle will be attracted to the regions of higher field intensity, *i.e.* positive DEP. On the other hand, if the particle has lower polarizability than the medium, the CM factor will be negative and the particle will be rejected from the regions of higher field intensity, *i.e.* negative DEP.

Additionally to DEP force, electrophoresis and electroosmotic flow are other electrokinetic forces present when DC electric fields are employed. The superposition of electrophoresis and electroosmotic flow is known as electrokinetic force (EK), which is lineal with the electric field. This suggests that there are two DEP regimes, one known as streaming DEP where DEP cannot overcome EK and particles flow freely from one electrode to the next, and trapping DEP where DEP is highly enough to overcome EK and particles are immobilize.

Experimental section

Sample preparation

Selenastrum capricornutum (UTEX 1648), a green algae, was cultured in 3N Bold medium, harvested at mid logarithmic phase and resuspended in bidistilled water. The sample was divided into two equal parts, one of which was heat-treated at 70 °C for 30 minutes to compromise viability of the cells. Afterwards the two samples were mixed to achieve 1:1 ratio of viable and non-viable algae and stained with Sytox® green dye (Invitrogen, Carlsbad, CA, USA), ex/em 504/523 nm, which can only penetrate cells with damaged membranes to intercalate with its DNA. The sample of viable cells was not stained, since the autofluorescence of chlorophyll A (ex/em 430/663), present in the strain was used to visually differentiate the cells from the heat-treated sample. The final sample was resuspended in a buffer with a pH of 9 and conductivity of 187 μS/cm to achieve a concentration of 1.7×10^7 cells/mL.

Microdevice description

A microchannel fabricated from glass was used for DEP experiments with cylindrical insulating posts straddled between two reservoirs. The channel is 10.16 mm-long, 2 mm-wide, and 20 μm-deep. The posts are arranged in four rows and eight columns and have a diameter of 440 μm and separated 520 μm center-to-center. The first and last columns have "dove-tail" geometry to prevent particle agglomeration by collision with the posts and in this way preventing plugging of the system. A schematic representation of the microdevice used is shown in Figure 2.

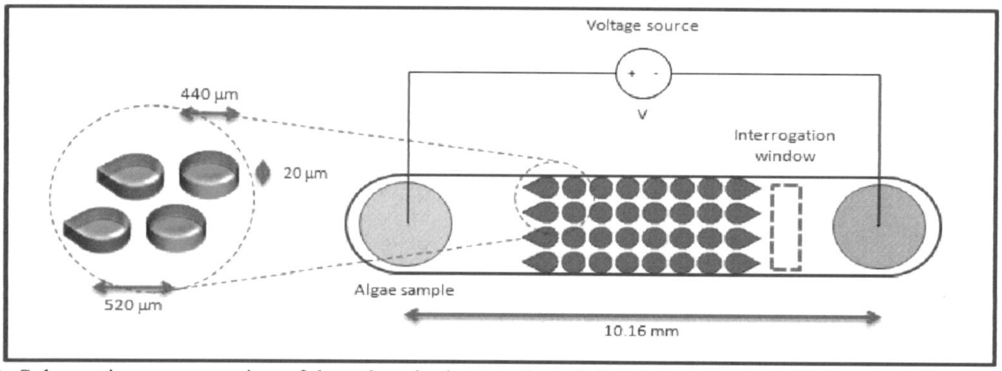

Figure 1. Schematic representation of the microdevice employed. Insulator posts are 440 mm in diameter and are separated 520 mm center-to-center. Sample is placed in the inlet reservoir and electrodes are positioned between the inlet and outlet reservoirs to induce electric fields.

Equipment

The dielectrophoretic behavior of the algae was visualized employing an inverted microscope model Axiovert 200 (Carl Zeiss, Göttingen, Germany), which has an integrated camera, and recorded as video utilizing the Microscope Software AxioVision LE (Carl Zeiss, Göttingen, Germany). A 10x microscope objective was used for all experiments. The microscope is equipped with an UV lamp that was used for simultaneous excitation of both fluorescent tags present in the sample: Sytox Green and chlorophyll A, allowing discrimination of viable (red) and non-viable (green) cells. For cell sizing, a digital microscope Motic® DM-B1 (Speed Fair Co., Ltd., Causeway Bay, Hong Kong) was used.

DC electric fields were applied by employing a 3000 V high voltage sequencer, model HVS448 (Labsmith, Livermore, CA, USA), using 0.3048-mm-diameter platinum-wire electrodes (Omega, Stamford, CT, USA) at each of the reservoirs. The microscopes and the high voltage sequencer require the use of a personal computer.

iDEP experiments

The microdevice described was first filled with buffer with controlled pH of 9 and conductivity of 187 μS/cm. Then a sample 50 μL of the 1:1 mixture of viable and non-viable microalgae was placed in the inlet reservoir and electrodes were positioned at the inlet and outlet reservoirs to induce DC electric fields.

Results and discussion

When a cell dies, its membrane is compromised, allowing for high conductive particles from the cell cytoplasm to penetrate membrane space and increase its conductivity. As mentioned, when a non-uniform electric field is applied to a sample of different particles, all cells will experience a dielectrophoretic force which magnitude will depend mainly on their conductivities and sizes. For a mixture of viable and non-viable algae cells of the same specie and culture, it is expected that size differences are negligible, and that differences in membrane conductivity are high enough to achieve dielectrophoretic differentiation.

Results show that when a low electric field is applied to the system, EK dominates and cells are directed to the outlet reservoir due to the negative net charge of the substrate of the microdevice. As mentioned in the theory section, EK is of first order to the electric field applied while DEP is of second order. Therefore, when the electric field is significantly increased, DEP will overcome EK and the particles will be immobilized (trapping DEP).[9] The magnitude of the electric field required to pass from the streaming regime to the trapping regime will be different for live and dead algae, due to their differences in cell membrane which directly affects the CM factor. Viable algae cells possess integrity in their membrane which will be reflected in lower conductivities, leading to higher magnitudes for the CM factor and hence higher DEP force.

Figure 2 presents a sequence of pictures obtained from a video taken with the equipment described in the experimental section. When an electric field of 1000 V/cm is applied the DEP force is strong enough to overcome EK for viable algae (red) but not for the non-viable algae (green). With the buffer used, pH 9 and conductivity of 187 μS/cm, both cells exhibit negative DEP and hence will be rejected from the regions of high field intensity (narrow regions between posts). This is seen as an accumulation of viable cells trapped before the posts while non-viable cells continue to flow dominated by EK (Fig. 2a). When the electric field is increased to 1700 V/cm the DEP force increases and overcomes EK for both cells and they are trapped but in different regions since DEP force is not equally exerted on them; repulsion from the narrow part between posts is stronger for live algae and therefore are trapped farther away (Fig. 2b).

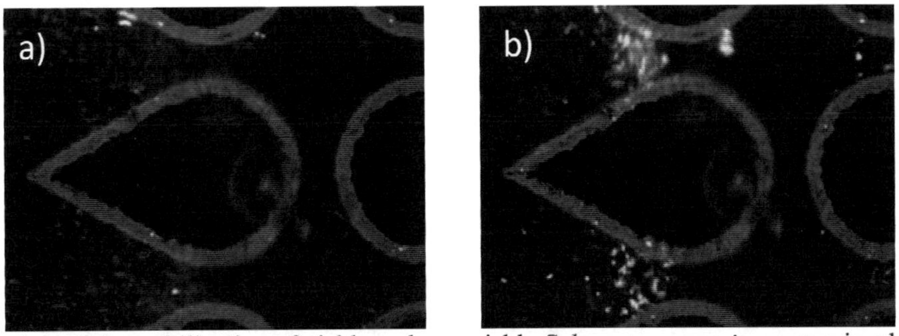

Figure 2. Dielectrophoretic immobilization of viable and non-viable *Selenastrum capricornutum* in a buffer of bidistilled water with a pH of 9 and a conductivity of 187 μS/cm. a) At E=1000 V/cm, only viable cells (red) are trapped by negative DEP while non-viable cells (green) are dominated by EK; b) At E= 1700 V/cm both cell types are trapped, but viable algae exhibit stronger repulsion from regions of higher field intensity (narrow regions between posts).

Conclusions

Separation between viable and heat-treated *Selenastrum capricornutum* was demonstrated in this work. This shows that viability of algae can be determined with simple iDEP experiments since live and dead cells experience different dielectrophoretic response under the same process circumstances. Differences in surface conductivities produce unequal CM magnitudes and therefore a separation of these two cells can be achieved.

Negative DEP was observed for both cells during this work since conductivity of algae is lower than that of the buffer used, this can be seen in the region where the algae was concentrated which is just before the region of high electric field intensity (narrow part between posts). The magnitude of the applied electric field can be manipulated to selectively trap or release only viable algae of both cells. The degree of the negative DEP trapping can be also controlled with the electric field.

Being iDEP a novel method for particle manipulation, there is a great potential to apply the technique for biotechnological procedures, *i.e.* viability of cells, as proved in this study.

References

1 USEPA, (1996), "Ecological effects test guidelines: OPPTS 850.5400 Algal toxicity, tiers I and II. EPA712-C-96-164.," U.S. Environmental Protection Agency, Office of Prevention, Pesticides and Toxic Substances, Washington, D.C.

2 Graham L.E., Wilcox L.W., (2000), "Algae," Prentice-Hall, Upper Saddle River, NJ.

3 Raja R., Hemaiswarya S., Kumar N.A., Sridhar S., Rengasamy R., (2008), "A perspective on the biotechnological potential of microalgae," *Crit. Rev. Microbiol.* 34, pp.77-88.

4 Pouneva I., (1997), "Evaluation of Algal Culture Viability and Physiological State by Fluorescent Microscopic Methods," *Bulg. J. Plant Physiol.* 23(1-2), pp.67-76.

5 Markelova A.G., Vladimirova M.G., Kuptsova E.S., (2000), "A comparison of cytochemical methods for the rapid evaluation of microalgal viability.," *Russ. J. Plant. Physiol.* 47, pp.815-819.

6 Capasso J.M., Cossío B.R., Berl T., C.J. R., Jiménez C., (2003), "A colorimetric assay for determination of cell viability in algal cultures," *Biolmol. Eng.* 20, pp.133-138.

7 Sato M., Murata Y., Mizusawa M., Iwahashi H., Oka S., (2004), "A Simple and Rapid Dual-fluorescence Viability Assay for Microalgae," *Microbiol. Cult. Coll.* 20(2), pp.53-59.

8 Li L., Song L., (2007), "Applicability of the MTT assay for measuring viability of cyanobacteria and algae, specifically for Microcystis aeruginosa (Chroococcales, Cyanobacteria)," *Phycologia* 46, pp.593-599.

9 Cummings E.B., (2003), "Streaming dielectrophoresis for continuous-flow microfluidic devices," *IEEE Eng. Med. Biol. Mag.* 22(6), pp.75-84.

(5c) Engineering Protein Electrophoresis through Surfactant Design: The Role of Nanostructure

Danilo C. Pozzo and Monica Ospinal, Chemical Engineering, University of Washington, Seattle, WA

Protein electrophoresis is a common method for the analysis and identification of specific proteins within complex mixtures (e.g. cell lyses). The surfactant that is utilized in the electrophoresis system plays an integral role in determining the resulting separation. The two main roles of the surfactant are to provide a uniform charge and also to control the hydrodynamic interactions of the denatured protein complex and the gel sieve. Since the introduction of polyacrylamide gel electrophoresis (PAGE), sodium dodecyl sulfate (SDS) has been utilized as the standard surfactant for separations. In this work we utilize small angle scattering techniques (SANS and SAXS) to probe the nanometer scale structure of protein-surfactant complexes that are formed with anionic surfactants of the SDS family but having variable tail architectures. Our aim is to correlate the nanostructure of the protein-surfactant complex to the quality and performance of the macroscopic separation so that optimized surfactant chemistries can be engineered using rational design rules.

See more of this Session: Advances in Electrokinetics and Electrophoresis: Fundamentals
See more of this Group/Topical: Topical 3: 2010 Annual Meeting of the American Electrophoresis Society (AES)

(5d) Investigation of Natural pH Gradient Formation within Microchannels

Aytug Gencoglu and Adrienne Minerick, Department of Chemical Engineering, Michigan Technological University, Houghton, MI

When an electric field is applied between an electrode pair in an aqueous medium, H+ and OH- ions are generated at anode and cathode surfaces, respectively. In micro- and nano-scale systems, the mass transport of these ions due to diffusion and electrokinetic transport can be significant enough that they develop concentration gradients across the system. The resulting "natural pH gradients" can be used for some techniques such as isoelectric focusing. On the other hand, natural pH gradients can complicate other systems by causing spatial changes such as nonuniform wall surface charges. The roles of diffusion, electrophoretic transport and transport due to electroosmotic flow were compared. COMSOL simulations were used to predict the length scales for each transport mechanism as a function of electric field strength, concentration in the electrode fluid wells, reaction rate at the electrode surface, as well as time scales needed for a pH gradient to develop down microchannels traversing the electrodes. The decay of established gradients after stoppage of the electric field was also simulated and experimentally observed in order to determine system relaxation times. To experimentally validate the simulation results, pH changes in a microchannel were measured by fluorescence microscopy. The pH-sensitive fluorescent dyes SNARF-1 dextran conjugate (SNARF-1) and FITC Isomer I (FITC) were used to measure pH at different locations and points in time during and after the application of an electric field, and the measured changes in the pH gradient were compared with the simulation results. This work has the potential to elucidate subtle ion movements in common electrokinetic microdevices such that processes can be engineered to avoid or take advantage of these natural ion gradients.

See more of this Session: Advances in Electrokinetics and Electrophoresis: Fundamentals
See more of this Group/Topical: Topical 3: 2010 Annual Meeting of the American Electrophoresis Society (AES)

(5e) Ion-Electron Transducing Electrodes with Electrochemically Active Conductive Material for Electroosmotic Pumps in Lab-On-A-Chip Applications

Per G. Erlandsson and Nathaniel D. Robinson, Dept. of Physics, Chemistry and Biology, Linköping University, Linköping, Sweden

Microfluidic Lab-on-a-chip (LOC) systems have the potential to revolutionize the healthcare industry by moving advanced diagnostic techniques into patients' homes and opening new possibilities in areas where healthcare infrastructure is limited. In order for LOC devices to reach their full potential they need to function independently of large external equipment. One of the key components to be integrated is the pump, which moves liquid samples between reactors, separators and detectors. Electroosmotic (EO) pumps drive plug-like flow and can easily be integrated in microfluidics systems, however, EO pumps usually produce gas bubbles (H_2 and O_2), hydrogen peroxide and/or acid and base at the electrodes. The hydrolysis of water changes the pH of the system, something which is unacceptable for microfluidic devices with sensitive materials (*e.g.* proteins) to be studied. By using electrodes based on electrochemically active conducting material functioning as ion-electron transducers, we have created EO pumps with significantly reduced hydrolysis and impact on pH, making these integrated pumps ideally suited for LOC applications.

The device described utilizes the ability of many semiconducting polymers and metal oxides to undergo electrochemical changes, and uses this ability to solve many significant challenges in the miniaturization of analytical biological and chemical systems into LOCs.

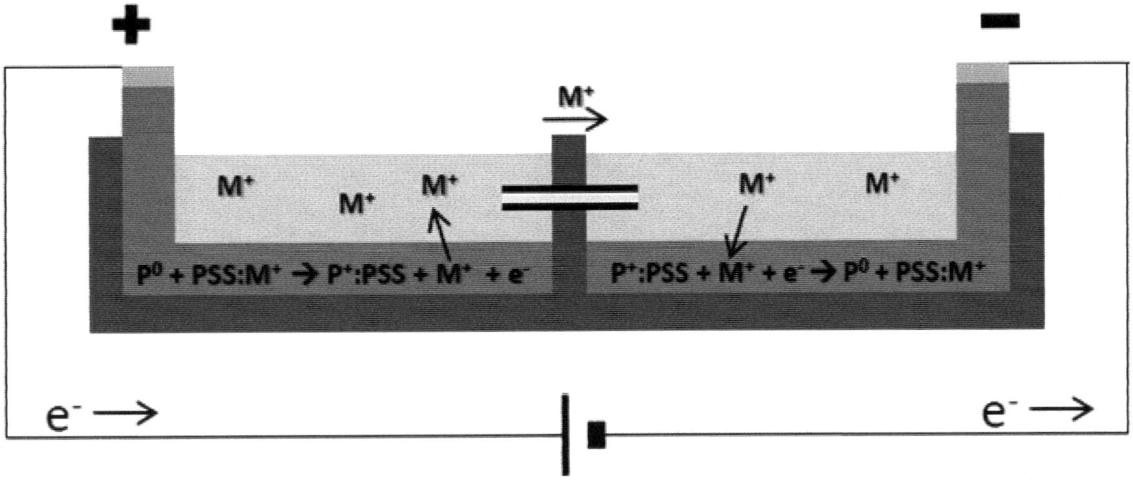

See more of this Session: Advances in Electrokinetics and Electrophoresis: Fundamentals
See more of this Group/Topical: Topical 3: 2010 Annual Meeting of the American Electrophoresis Society (AES)

(5f) Dynamics of Separation in An Electrical Field Flow Fractionation (EFFF) Separator with Couette Flow: Effect of Wall Velocities

Jennifer Anne Pascal[1], Mario Oyanader[2], Pedro Arce[2] and Yung Way Liu[3], (1)Chemical Engineering, Tennessee Tech University, Cookeville, TN, (2)Chemical Engineering, Tennessee Technological University, Cookeville, TN, (3)Mathematics, Tennessee Technological University, Cookeville, TN

Improving the separation of biomacromolecules, such as proteins, DNA, and antibiotics, to name a few, can lead to the development of novel pharmaceuticals. Electrokinetic-based methods, including electrical field flow fractionation can be used to effectively separate a variety of molecules. In electrical field flow fractionation (EFFF), an electrical field is applied perpendicular to the direction of flow in a channel, and the molecules most susceptible to the electrical field will migrate fastest to the walls, thus creating a very efficient separation (Giddings, 1993).

In this contribution, the dynamics of an EFFF separator with Couette flow is examined. In the system under study, both walls of the device move at nonzero velocities and in opposite directions to produce a Couette type of flow. Therefore, it is useful to study the effects of the wall velocities on the dynamics of separation within the device. This is accomplished by solving the transient averaged species continuity equation analytically, to obtain average concentration profiles as a function of various parameters of the system, including, the applied orthogonal electrical field, valence, electrophoretic mobility, and wall velocities. This analysis enables the determination of optimal operating conditions of the device under transient conditions.

See more of this Session: Advances in Electrokinetics and Electrophoresis: Fundamentals
See more of this Group/Topical: Topical 3: 2010 Annual Meeting of the American Electrophoresis Society (AES)

(5g) Super-Limiting Current in Porous Media

E. Victoria Dydek, Klavs F. Jensen and Martin Z. Bazant, Chemical Engineering, Massachusetts Institute of Technology, Cambridge, MA

In classical electrochemical systems, the maximum sustainable current to a membrane or electrode is limited by diffusion and convection. In this work, we show that super-limiting currents can be achieved in micro-pores due to the nonlinear effect of surface conduction, even in the absence of convection. While the neutral bulk current remains limited, the surface current through the diffuse double layers allows for higher total currents, even at modest potentials. This phenomenon generates strong concentration polarization and depletion regions that depend on two dimensionless parameters: the Dukhin number, Du (ratio of surface and bulk conductivities), and Peclet number, Pe (ratio of convection and diffusion lengths). Although the initial double layers are thin with Du<<1, once the limiting current is exceeded, the depletion region itself will exhibit strong surface effects with Du=O(1). The appearance of these depletion regions suggests that judicious application of surface conduction can be used for electrodialysis. Energy/volume calculations suggest that this method would have efficiencies of the same order as reverse osmosis (10kWhr/m3.) Theoretical discussions with numerical results are presented and supported with experimental data.

See more of this Session: Advances in Electrokinetics and Electrophoresis: Fundamentals
See more of this Group/Topical: Topical 3: 2010 Annual Meeting of the American Electrophoresis Society (AES)

(5h) A Microfluidic Platform for Enhancing Quantitation of Biological Mixtures

Tristan F. Kinde and Debashis Dutta, Chemistry, University of Wyoming, Laramie, WY

In any analytical method of biomarker identification, great lengths must be made to extract or purify specific molecules prior to instrumental detection. In the case of proteomic analysis of a cell lysate via mass spectrometry, for example, counter ions (DNA) and neutral molecules (lipids and triglycerides) can inhibit signal and increase systematic noise, increasing the limit of detection. The signal inhibition is likely caused by the electrostatic coupling of the proteins/peptides with anions (DNA) or micelle formation with non-polar molecules (lipids). Using a microfluidic device with integrated electrodes, we have explored and optimized methods to create enriched biomolecule fractions for such proteomic analyses based on the Electric Field Flow Fractionation (EFFF) technique. With a simple mixture of DNA and peptide, we have demonstrated that over 95% of the peptide can be separated from the DNA with a molecular throughput of about 1.2x10-9 mol/min. Further, we have observed an improvement in the limit of detection for peptides by nearly an order of magnitude when these anionic components (DNA) are removed from the sample using the reported microfluidic charge-induced lateral separation platform integrated to an electrospray ionization source (ESI-MS). Currently, we are evaluating the performance of this microfluidic device as a sample preparation unit for improving the inline quantitation of various biological mixtures.

See more of this Session: Advances in Electrokinetics and Electrophoresis: Fundamentals
See more of this Group/Topical: Topical 3: 2010 Annual Meeting of the American Electrophoresis Society (AES)

(5i) An Induced-Charge Electro-Osmosis Micropump

Joel S. Paustian, Chemical Engineering, University of California, Santa Barbara, Santa Barbara, CA and Todd Squires, Chemical Engineering, University of California Santa Barbara, Santa Barbara, CA

Microfluidic devices (e.g. Labs on a Chip) are becoming useful scientific and medical tools for automating chemical and biological lab work. Various impediments prevent complex microfluidic devices from being easily removed from a laboratory setting, limiting their utility for day-to-day applications like in-the-field medical diagnostics and drug delivery. The development of portable and integrable high-pressure pumping techniques will be necessary step for truly portable, complex microfluidic devices. Microfluidic pumps based on the electrokinetic phenomenon of Induced-Charge Electro-Osmosis (ICEO) could potentially fill this role. We describe ICEO and present a simple idea for a low-voltage, high-pressure micropump. We give simple scaling arguments, and a detailed theory, for its expected performance, and describe the design, fabrication, testing and characterization of a functional ICEO micropump. Our results validate the central idea, are consistent with our theoretical expectations, and suggest routes for the optimization and eventual use of the pump

See more of this Session: Advances in Electrokinetics and Electrophoresis: Fundamentals
See more of this Group/Topical: Topical 3: 2010 Annual Meeting of the American Electrophoresis Society (AES)

(5j) Analysis On the Effect of Channel Morphology for Electrophoresis of Bio-Molecules: Dynamics

Jyothirmai J. Simhadri, Mario Oyanader, Holly Stretz and Pedro Arce, Chemical Engineering, Tennessee Technological University, Cookeville, TN

Gel electrophoresis is the most widely used technique for the separation and purification of biomolecules (DNA and proteins). The nanometer-scale pore structure of the gel tuned either by templating or by adding nanoparticles has been shown to alter the electrophoretic mobility of the biomolecules thus improving separation efficiency. Moreover, analysis of the effects of the pore length and diameters on the transport of biomolecules is an important aspect to be studied either analytically or computationally as shown by previous efforts [1, 2]. In this research, we propose to computationally analyze different pore models (associated with gel materials) and study the effect of geometry on the transport properties in this anisotropic-like media. Results will illustrate, for example, the role of pore-scale in conjunction with the effect of secondary electrical fields.

References: 1. Trinh, S., B.R. Locke and P.E. Arce, Separation and Purification Technology, 15, 255 (1999). 2. Hidalgo, R., M. A. Oyanader (*), and P. E. Arce, AIChE Annual Meeting, Salt Lake City, 2007. (*) Speaker.

See more of this Session: Advances in Electrokinetics and Electrophoresis: Fundamentals
See more of this Group/Topical: Topical 3: 2010 Annual Meeting of the American Electrophoresis Society (AES)

(96a) Desalination Shocks in Microstructures

Ali Mani, Mechanical Engineering, Stanford University, Stanford, CA, E. Victoria Dydek, Chemical Engineering, Massachusetts Institute of Technology, Cambridge, MA and **Martin Z. Bazant**, Chemical Engineering and Mathematics, Massachusetts Institute of Technology, Cambridge, MA

Transport in quasi-neutral bulk electrolytes is governed by linear diffusion and convection. As first noted by J. J. Bikerman in the 1930s, the proximity of charged walls increases the importance of surface conduction and electro-osmotic flow, driven by diffuse ionic charge. In nano-channels and membranes with overlapping double layers, these interfacial effects dominate diffusion, resulting in linear electrokinetic phenomena. In microchannels and porous media, however, bulk and interfacial transport compete with each other, and the dynamics of concentration polarization become nonlinear. Mani, Zangle and Santiago recently showed that nearly discontinuous shocks in the conductivity can propagate from a microchannel/nanochannel junction, analogous to shock waves in gases, even if the double layers are thin. This talk describes the basic physics of such "desalination shocks" and develops a mathematical theory of concentration polarization in non-uniform microstructures or porous media. The theory predicts a new mechanism for super-limiting current to membranes or electrodes, driven by surface conduction, even in the absence of convection.

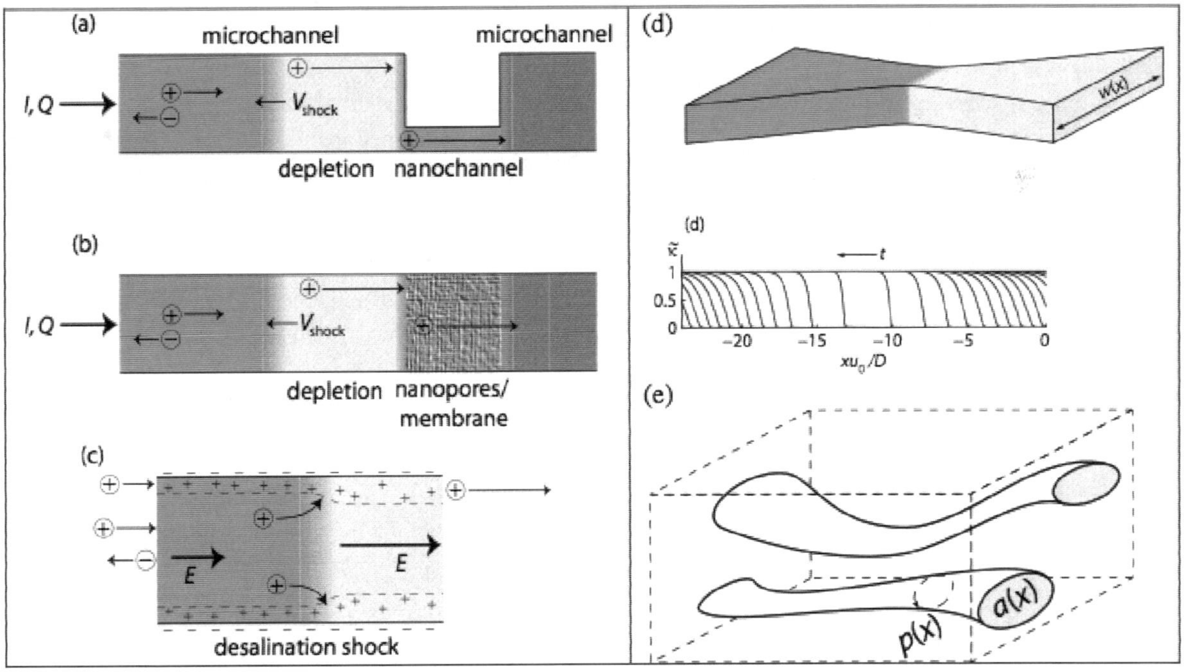

See more of this Session: Fundamentals of Electrokinetic Flows: Novel Applications and Ionic Fluxes at Interfaces
See more of this Group/Topical: Topical 3: 2010 Annual Meeting of the American Electrophoresis Society (AES)

(96b) Real-Time PCR Quantification by Asymmetric Electrophoresis and Warburg Impedance Across Nanoslots

Sagnik Basuray[1], Peter Mushenheim[1], Thomas Hagan[1], Satyajyoti Senapati[1], Gilad Yossifon[2] and H.-C. Chang[1], (1)Chemical and Biomolecular Engineering, University of Notre Dame, Notre Dame, IN, (2) Mechanical Engineering, Technion, Technion City, Israel

The development of portable, inexpensive, label-free and accurate platform for the quantification of DNA concentration is highly desirable, especially for real-time monitoring of the number of copies of DNA amplified in PCR. A novel nano-slot sensor design is proposed that employs an electrochemical impedance meter and a small AC field to quantify DNA concentration with pico-molar sensitivity. The DNA solution is introduced into two micro-reservoirs connected by a nano-slot across which a low-frequency AC field is applied. The cross-channel impedance spectrum of the system varies with time, as the co-ion DNA molecules are driven to the nano-slot entrance by the AC field. This trapping mechanism is shown to be a result of asymmetric electrophoresis due to enrichment/depletion symmetry breaking across the nanoslot. The larger voltage drop at the depletion region reduces its bulk field, resulting in a smaller bulk electrophoretic displacement during the depletion half cycle. The high field and ion depletion force in the depletion region also combines with the dominant dielectrophoretic force to endow a size-sensitive concentration rate. To validate this theory, microscopy experiments are performed using fluorescently labeled DNA to visualize the asymptotic electrophoresis of DNA at the nano slot. Due to the high charge of the DNA macro-ions and its large associated counter-ions, their presence in the depletion layer introduces a surfeit of counter-ions that significantly increases the conductivity of the depletion layer. Since the depletion layer is the region with the lowest conductivity in the entire device, the presence of DNA molecules significantly changes the overall conductance of the nano-slot system. This conductance change appears as a shift in the low-frequency impedance spectrum. The magnitude of this shift is directly correlated with the concentration of DNA present in the system. This allows for a highly sensitive, rapid, label-free and inexpensive real-time DNA quantification device for point-of-care application.

>
See more of this Session: Fundamentals of Electrokinetic Flows: Novel Applications and Ionic Fluxes at Interfaces
See more of this Group/Topical: Topical 3: 2010 Annual Meeting of the American Electrophoresis Society (AES)

(96c) Electrokinetic Flows in Gradient Monoliths

Cornelius F. Ivory, Voiland School of Chemical Engineering and Bioengineering, Washington State University, Pullman, WA

When a pressure gradient is applied across an open capillary, the resulting parabolic flow causes a dramatic dispersion of solutes in that tube. If the same capillary is packed with a monolith or granular stationary phase, then applied pressure and electric fields will generate uniform flows with very low dispersion. However, when electric fields are applied across stationary phases with gradients in fixed charges, the resulting flows can exhibit anomalous behavior, such as flow reversal, which can lead to pathological dispersion during electrofocusing. Because the stationary phase nanopores are roughly the same scale as the Debye length, flows through the mobile and stationary phase are a strong function of charge density, buffer conductivity, and pore size. Because of this, it should be possible to optimize the flow properties for a given separation by adjusting the ionic strength of the running buffer.

This paper will begin with a mathematical model for combined hydraulic- and electrokinetically-driven flows in gradient monoliths and then use that model to predict how these flows vary with the properties of the monoliths and the running buffer. The model will then be used to predict electrofocusing performance on gradient monoliths.

See more of this Session: Fundamentals of Electrokinetic Flows: Novel Applications and Ionic Fluxes at Interfaces
See more of this Group/Topical: Topical 3: 2010 Annual Meeting of the American Electrophoresis Society (AES)

(96d) DNA Electrophoresis through Nanopores

Sandip Ghosal, Mechanical Engineering, Northwestern University, Evanston, IL

The voltage induced translocation of linear polyelectrolytes (such as DNA) through nanometer scale pores may be used as a basis for size separation. With further refinement it might be possible to extract more detailed information such as the sequence of bases from such experiments. The observables are the change in the magnitude and temporal duration of pore conductance during an individual translocation event. It is shown that these quantities may be reliably calculated by modeling a translocation event in terms of the electrophoresis of a charged cylinder along the axis of a cylindrically symmetric pore. Theoretical calculations are compared to data from single molecule experiments involving free translocations as well as force measurements using DNA immobilized within a nanopore by tethering it to a bead held in an optical trap.

See more of this Session: Fundamentals of Electrokinetic Flows: Novel Applications and Ionic Fluxes at Interfaces
See more of this Group/Topical: Topical 3: 2010 Annual Meeting of the American Electrophoresis Society (AES)

(96e) Length Dependent Electrophoretic Mobility of Polyelectrolytes in Uniform Shear Flow

Rahul Kekre, J. E. Butler and A.J.C. Ladd, Department of Chemical Engineering, University of Florida, Gainesville, FL

Recent work suggests that hydrodynamic interactions generated by an electric field acting on a polyelectrolyte are not fully screened. As a result the electrophorectic mobility of an elongated polyelectrolyte may contain additional contributions from interactions between distant parts of the chain. We have developed a Brownian-dynamics simulation that includes electrically driven hydrodynamic interactions, which are dipolar in nature in low-salt conditions. In the absence of flow, the electrophoretic mobility is independent of the chain length as the polymer retains a spherical shape. However, in a uniform shear flow we find that the electrophoretic mobility increases with shear rate, initially linearly but later saturating at a value approximately 10% higher than the mobility of the undeformed chain. At the same flow rate, the electrophoretic mobility of a sheared polyelectrolyte varies with the contour length.

See more of this Session: Fundamentals of Electrokinetic Flows: Novel Applications and Ionic Fluxes at Interfaces
See more of this Group/Topical: Topical 3: 2010 Annual Meeting of the American Electrophoresis Society (AES)

(96f) Understanding Protein-Surfactant Structure for Efficient and High Resolution Electrophoretic Separation On Microchip Platforms

Glareh Azadi, Engineering, Brown University, Providence, RI, Anuj Chauhan, Chemical Engineering, University of Florida, Gainesville, FL and Anubhav Tripathi, Brown University Division of Engineering, Providence, RI

The study of interactions of protein with surfactants is not only important from biochemical stand point to determining the mechanism, but also is necessary in many applications such as pharmaceutical, cosmetic and food industry. In recent years, microfluidic devices have been demonstrated as effective analytical tools in the field of protein separation, detection and quantification. Here we present interaction of proteins with both anionic and cationic surfactants in presence of hydrophobic dye molecules. Using fluorescent microscopy, we quantify these interactions for number of proteins. The results of this study can be used to develop new methods for sizing, sensitive detection and quantification of proteins and protein isoforms. We will also present microfluidic separation of proteins under different conditions of channel surfaces. The results will be discussed using the fundamental principles of protein migration in polymer gels and transport equation.

See more of this Session: Fundamentals of Electrokinetic Flows: Novel Applications and Ionic Fluxes at Interfaces
See more of this Group/Topical: Topical 3: 2010 Annual Meeting of the American Electrophoresis Society (AES)

(96g) 2D Assembly of Colloid Particles On An Electrode within the Ideally Polarizable Region

Christopher L. Wirth, Paul J. Sides and Dennis C. Prieve, Department of Chemical Engineering and Center for Complex Fluids Engineering, Carnegie Mellon University, Pittsburgh, PA

A low frequency (<1kHz) AC electric field induces particle assembly on an electrode in the plane perpendicular to the electric field's direction. The dispersing electrolyte has been found to have a profound effect on the direction of motion: particles aggregate in solutions of $NaHCO_3$ or KCl and move apart in solutions of KOH, NaOH, H_2CO_3 or NH_4OH. The direction of lateral motion between pairs correlates with the sign of $\cos\Theta$ measured on single particles, where Θ is the phase angle between the oscillations in the elevation of the particle and the oscillations in the electric field driving it. Previously, we speculated that an electrochemical reaction (producing faradaic current) occurring at low-frequencies was the reason for the apparent electrolyte dependence. Using a new tool, called the Electrochemical Total Internal Reflection Microscope (ETIRM), we identify regions where the working electrode is ideally polarizable (no possibility of faradaic current). We present data from single and multi-particle experiments with and without the possibility of faradaic current that suggest faradaically coupled electroosmosis plays little to no role in the electrolyte dependent assembly of particles in low frequency AC fields. An insight into the dynamics of particle motion at low frequencies leads to a new understanding of how Θ arises.

See more of this Session: Fundamentals of Electrokinetic Flows: Novel Applications and Ionic Fluxes at Interfaces
See more of this Group/Topical: Topical 3: 2010 Annual Meeting of the American Electrophoresis Society (AES)

(96h) Chemo and Phototactic Nano/Microbots

Ayusman Sen, Department of Chemistry, The Pennsylvania State University, University Park, PA

One of the most interesting scientific challenges is the design of synthetic motors on the nano and micrometer scales which could be used for drug delivery, self-assembly of superstructures, roving sensors, and other emerging applications. Our group was the first to report purely synthetic nano/ micromotors. These autonomous motors consist of bimetallic segments which utilize hydrogen peroxide to power their movement through the solution. The mechanism of motion was determined to be self-electrophoresis, in which redox reactions catalyzed at the two metals create an electric field in solution which acts back on the rod.

We have achieved the attachment, transport, and delivery of a prototypical cargo, polymer microspheres, using bimetallic nanomotors. This may find application in bottom-up assembly of colloids or for delivery of materials at a specific location whereupon further binding events may be triggered depending on cargo surface functionality. Finally, an object that moves by generating a continuous surface force in a fluid can, in principle, be used to pump the fluid by the same catalytic mechanism. Thus, by immobilizing these bimetallic nanomotors, we have developed micro/nanofluidic pumps that transduce energy catalytically.

We have also explored an alternative mechanism to power synthetic micro/nanomotors, self-diffusiophoresis. In our case, each particle generates its own gradient of electrolyte that then acts back on the particle. We have shown that particles that release ions under UV illumination exhibit the phenomenon of phototaxis, the first examples outside living systems. Further, the gradients produced by adjacent particles interact and the particles began to swarm into concentrated groups. In this respect, the particles are reminiscent of quorum sensing organisms that swarm in response to a minimum threshold concentration of a signaling chemical.

See more of this Session: Fundamentals of Electrokinetic Flows: Novel Applications and Ionic Fluxes at Interfaces
See more of this Group/Topical: Topical 3: 2010 Annual Meeting of the American Electrophoresis Society (AES)

(96i) Guided Motion of Self-Propelled Magnetic Colloidal Particles by Brownian Dynamics Simulations

Glenn Vidal, Carlos Rinaldi and **Ubaldo M. Córdova-Figueroa**, Department of Chemical Engineering, University of Puerto Rico–Mayagüez, Mayagüez, PR

Self-propulsion of artificial nano- and microscale objects by the transformation of chemical energy into motion is one of the most fascinating and exciting challenges currently studied. It has been shown in recent experiments that autonomous motion of the so-called 'catalytic' motors is hindered by their rotary Brownian motion and thus preventing its potential to be fully realized. However, such limitation could be relaxed with colloidal particles sensitive to external magnetic fields. The present study investigates the short and long-time diffusive behavior of a catalytically driven 'magnetic' colloidal particle immersed in a dispersion of reactant particles subject to a magnetic field using Brownian dynamics simulations. The strength of the magnetic field is controlled by the Langevin parameter, which physically measures the relative importance of magnetic to Brownian torques, and dictates the spatiotemporal behavior of the particle. The translational self-diffusivity is measured for different surface reaction speeds, particle sizes, reactant particle concentrations, magnetic dipole orientations, and Langevin parameters. Finally, a theory to determine the long-time self-diffusivity and time-averaged particle velocity is constructed and compared to the simulation results.

See more of this Session: Fundamentals of Electrokinetic Flows: Novel Applications and Ionic Fluxes at Interfaces
See more of this Group/Topical: Topical 3: 2010 Annual Meeting of the American Electrophoresis Society (AES)

(166a) Electrokinetic Locomotion by Reaction Induced Charge Auto-Electrophoresis

Jonathan D. Posner, Mechanical and Aerospace Engineering, Arizona State University, Tempe, AZ and Jeffery L. Moran, Mechanical Engineering, Arizona State University, Tempe, AZ

The controlled motion of synthetic nanoscale motors may represent a major step towards the development of practical nanomachines, artificial cells, and autonomous microsystems. We are investigating locomotion of bimetallic synthetic nanomotors that, analogous to their biological counterparts, harvest chemical energy from their local environment and convert it to useful work. Bimetallic nanorods can autonomously propel themselves at a hundred body lengths per second through aqueous solutions by using hydrogen peroxide as a fuel. We can control the motion of nanomotors using magnetic fields and chemical species to control the motion of Pt-Ni-Au nanorods.

Several arguments have been proposed to describe the physics underlying their locomotion, but there is no detailed theory on the propulsion mechanism. We are simulating the physics of rod-shaped nanoparticles with asymmetric surface fluxes. Our models show that locomotion is driven by electric body forces in the fluid that arise due to finite space charge and internally generated electric fields surrounding the rod. The electric fields and charge density are generated by dipolar cation fluxes, such as those generated by heterogeneous electrochemical reactions with broken symmetry. We present a set of governing equations, a scaling analysis, numerical simulations, and experiments that describe the physics underlying the autonomous motion of electrocatalytic bimetallic nanomotors due to a mechanism we call Reaction Induced Charge Auto-Electrophoresis (RICA).

See more of this Session: Plenary Session of the American Electrophoresis Society
See more of this Group/Topical: Topical 3: 2010 Annual Meeting of the American Electrophoresis Society (AES)

(166b) Non-Coalescence of Oppositely Charged Drops

William D. Ristenpart, Dept. of Chemical Engineering & Materials Science, University of California, Davis, Davis, CA

Electrically induced droplet motion manifests itself in processes as diverse as storm cloud formation, commercial ink-jet printing, petroleum and vegetable oil dehydration, electrospray ionization for use in mass spectrometry, electrowetting and lab-on-a-chip manipulations. An important issue in practical applications is the tendency for adjacent drops to coalesce, and oppositely charged drops have long been assumed to experience an attractive force that favors their coalescence. Here we describe the existence of a critical field strength above which oppositely charged drops do not coalesce [1]. We observe that appropriately positioned and oppositely charged drops migrate towards one another in an applied electric field; but whereas the drops coalesce as expected at low field strengths, they are repelled from one another after contact at higher field strengths. Qualitatively, the drops appear to 'bounce' off one another. We directly image the transient formation of a meniscus bridge between the bouncing drops, and we propose that this temporary bridge is unstable with respect to capillary pressure when it forms in an electric field exceeding a critical strength. Finally, we establish techniques for exploiting the phenomenon to tune the trajectory and partial coalescence of oppositely charged drops to obtain drops in desired locations or of a desired size. The observation of oppositely charged drops bouncing rather than coalescing in strong electric fields should affect our understanding of any process involving charged liquid drops, including de-emulsification, electrospray ionization and atmospheric conduction.

[1] Ristenpart et al., Nature 461, 377-380 (2009).

See more of this Session: Plenary Session of the American Electrophoresis Society
See more of this Group/Topical: Topical 3: 2010 Annual Meeting of the American Electrophoresis Society (AES)

(166c) Ion Conservation and Its Consequences for Electro-Osmosis and Electrophoresis

Todd M. Squires, Chemical Engineering, University of California, Santa Barbara, Santa Barbara, CA

I will discuss electrokinetic flows, in which applied electric fields exert forces upon ionic clouds that surround charged surfaces, forcing the surrounding fluids into motion. Such effects have been known for two centuries, and a rational theory has been around for one. Given this history, one has no right to expect new surprises to emerge. Developments in micro- and nano-fluidic systems, however, have subjected the classic theory to new tests, and challenged our understanding of even the most basic concepts in this arena. I will describe cases in which the "standard model" of coupled ion transport and viscous flows are inadequate to explain measured phenomena, and will describe various mechanisms by which such discrepancies might arise. In particular, I will describe cases where non-uniform ion transport along surfaces -- e.g. nanoscale roughness, variable surface charge, or inhomogeneous slip -- changes even the qualitative features of electrokinetic flows.

See more of this Session: Plenary Session of the American Electrophoresis Society
See more of this Group/Topical: Topical 3: 2010 Annual Meeting of the American Electrophoresis Society (AES)

(166d) A New Quantitative Molecular Detection Platform for Field Applications

Hsueh-Chia Chang, Department of Chemical and Biomolecular Engineering, University of Notre Dame, Notre Dame, IN

Despite claims to the contrary, commercially viable field-use genetic or immuno-assays have yet to appear on the market, even though upstream units like PCR chips, cell-lysing units and bead/membrane filtration technologies have matured sufficiently to be integrated into a hand-held portable device. The key technology bottle neck is detection. Current technologies, like real-time PCR, rely on fluorescent optical detection, which requires bulky and expensive equipment unsuitable for field applications. Label-free electrochemical sensors for DNA, RNA and peptide biomarkers, with functionalized molecular probes, are believed to be the most promising detection platform for such portable units. However, classical electrochemical sensing technologies based on conductance, capacitance or reactance measurements remain very insensitive, even with reporter redox pairs, voltage-gating and surface monolayer (SAM) electrode modification. Also, like fluorescent sensing, their diffusion-limited assay time is in excess of one hour for a typical field sample volume. We report here several field-use enabling breakthroughs in our laboratory for electrochemical sensing. A flow format is designed to effect convection-enhanced mass transfer and target concentration. The high flow rate also produces a shear-enhanced selectivity unavailable in standard batch optical platforms. A low-frequency AC field is applied across a nanoporous structure/membrane to deionize the flowing sample near the electrode sensor to enhance sensitivity by orders of magnitude. The high-field in the deionized region allows dielectrophoretic trapping of large target molecules from the flowing solution. The same AC field produces a Warburg spectrum at the electrode sensor to allow sensitive and quantitative detection of local conductivity/reactivity changes due to the presence of only a few thousand captured molecules. The result is a high throughput (10 microliter/min), high selectivity (SNP discrimination), low detection limit (picoMolar), calibrated (picoM to microM dynamic range) and rapid (10 minutes) detection platform that is currently being integrated with upstream units to produce a turn-key and portable commercial device that includes debris filtration, lysing and PCR amplification.

See more of this Session: Plenary Session of the American Electrophoresis Society
See more of this Group/Topical: Topical 3: 2010 Annual Meeting of the American Electrophoresis Society (AES)

(166e) A Microfluidic Platform for Studying Host-Pathogen Interactions with Single-Cell Resolution

Anup K. Singh, Sandia National Laboratories, Livermore, CA

Innate immunity is our first line of defense against a pathogenic bacteria or virus. A comprehensive "system-level" understanding of innate immunity pathways is the key to deciphering mechanisms of pathogenesis and may lead to improvements in early diagnosis or developing improved therapeutics. Current methods for studying signaling focus on measurements of a limited number of components in a pathway and hence, fail to provide a systems-level understanding. Current methods also lack single-cell resolution necessary for accurate quantification of concentration and kinetics. We have developed a systems biology approach to decipher toll-like receptor 4 (TLR4) pathways in murine macrophages in response to exposure to pathogenic bacteria and their lipopolysaccharide (LPS). Our approach integrates a microfluidic cell handling and analysis platform with high-resolution imaging to provide spatially- and temporally-resolved measurement of signaling pathways. The integrated microfluidic platform is capable of imaging single cells to obtain dynamic translocation data as well as high-throughput acquisition of quantitative protein expression and phosphorylation information of selected cell populations. The platform consists of multiple modules such as single-cell array, cell sorter, and phosphoflow chip to provide confocal imaging, cell sorting, and flow cytometry-based phosphorylation assays. The single-cell array module contains fluidic constrictions designed to trap and monitor up to 100 single cells for hours (cells viable for >72hrs), enabling detailed statistically-significant measurements. The module was used to analyze translocation behavior of transcription factor NF-kB in macrophages upon activation by E. coli and Y. pestis LPS. Activation of NF-kB is preceded by phosphorylation of many kinases and to correlate the kinase activity with translocation, we performed flow cytometry-based phosphorylation assays in the PhosphoChip module. To allow further downstream analysis on selected cells, we also implemented an optical-trapping based sorting of cells. This has allowed us to sort macrophages infected with bacteria from uninfected cells with the goal of obtaining data only on the infected (the desired) population. The various microfluidic chip modules have been integrated into a portable package that can be mounted on a typical inverted microscope. The microfluidic platform allows high-resolution imaging as well as quantitative proteomic measurements with high sensitivity (

See more of this Session: Plenary Session of the American Electrophoresis Society
See more of this Group/Topical: Topical 3: 2010 Annual Meeting of the American Electrophoresis Society (AES)

(198a) Subcellular Fractionation in a Fluidic Microsystem by Dielectrophoresis (DEP)

Anton Posch[1], **Aran Paulus**[2], Monika Hausmann[3], Nancy Kunz[1], Martin Stelzle[3], Thanh Tu Duong[4] and Gert Blankenstein[4], (1)Bio-Rad Laboratories GmbH, Munich, Germany, (2)Bio-Rad Laboratories, Hercules, CA, (3)Natural and Medical Sciences Institute NMI, Reutlingen, Germany, (4)Boehringer Ingelheim microParts GmbH, Dortmund, Germany

In Proteome analysis the application of pre-fractionation techniques is a widely used strategy to reduce sample complexity and to enrich low abundance proteins. In this context, subcellular organelle fractionation is of particular importance, since functional units are isolated providing additional information with respect to the cellular localization of a particular protein (1, 2). Techniques currently employed for subcellular fractionation lack specificity to provide sufficiently pure organelle preparations for specific down-stream applications like comparative Proteome research. In order to overcome the above mentioned drawbacks of current organelle isolation techniques we have developed a micro fluidic system with integrated electrode arrays and a dedicated high through-put channel layout to isolate mitochondria from crude organelle samples based on dielectrophoretic sorting.

Mitochondria of a human lymphoblastoid cell line were stained with the fluorescent dye JC-1. Cells were homogenized with a douncer and the resulting cell homogenate was centrifuged at 1000xg for 10 min to remove cell debris and unbroken cells. The supernatant was centrifuged at 10.000xg for 10 min in order to obtain a crude mitochondrial preparation. The resulting pellet was suspended in a standard mitochondrial storage buffer at a concentration of 0.2 µg/µL and pumped through the microsystem at rates of 1 µl/min (Figure 1). Mitochondria were deflected by dielectrophoretic forces for 36 hours. Mitochondrial samples (before and after dielectrophoresis) were characterized either by 1-D electrophoresis followed by immunoblotting with organelle specific antibodies or by 2-D electrophoresis followed by mass spectrometric analysis.

Micro channels which contain deflector electrode arrays are generating inhomogeneous electric fields. Particles like cell organelles moving through the channels become polarized and experience dielectrophoretic forces whose magnitude depends on their size and dielectric properties. The small size of organelles of typically less than 1 µm in diameter require both high field strength and large degree of field inhomogeneity and consequently small electrode distances in the order of 10 – 20 µm. By variation of buffer flow velocity, voltage amplitude and frequency, optimum parameters for manipulation of mitochondria were determined. Deflection of mitochondria at 400 kHz and 4kV/cm was successfully demonstrated. In a time frame of 20 hours ~100 µg mitochondrial protein was collected and further analyzed by 1-D immunoblotting with organelle specific antibodies. Data evaluation of the immunoblots showed that the ratio between antibody signal intensity and protein amount loaded per lane differs significantly between the crude starting material and DEP sample. For VDAC (mitochondrial marker protein) an enrichment factor of about 2.5 can be calculated. Concerning BiP (marker protein for ER) a sixfold decrease in signal intensity is obvious for the DEP sample. LAMP1 (lysosomal marker protein) immunoblot signals were not evaluated in a quantitative manner, but almost no LAMP1-signal was observed for the DEP sample. 2D-PAGE analysis followed by mass spectrometry confirmed the superior purity of the DEP sample in comparison to standard gradient centrifugation techniques for the isolation of mitochondria.

References (1) H. Lu et al., Anal.Chem. 76, 5705-5712 (2004). (2) S. Brunet et al., Trends in Biotechnology 13, 829-637 (2003).

See more of this Session: Advances in Proteomics: New Technologies
See more of this Group/Topical: Topical 3: 2010 Annual Meeting of the American Electrophoresis Society (AES)

(198b) A Microscale Platform for Integrated Cell-Free Expression and Activity Screening of Cellulases

Aarthi Chandrasekaran, Rajiv Bharadwaj, Joshua I Park, Rajat Sapra, Masood Z Hadi, Paul D Adams and Anup K Singh, Joint BioEnergy Institute, Emeryville, CA

Cellulases play a key role in the bioconversion of lignocellulosic material to ethanol by hydrolyzing the cellulose polymers in biomass to fermentable sugars. However, the high cost of cellulases limits their utility in commercial biorefineries. Several research efforts have therefore been dedicated to the identification and generation of cost-effective, improved cellulases for the efficient hydrolysis of cellulose. The success of these efforts relies on the development of high-throughput integrated analytical approaches for rapid and efficient expression and screening of cellulases. To address this need, we have developed a miniaturized, fluorescence-based platform that integrates cell-free expression and activity profiling of cellulases. Our approach achieves rapid cellulase expression and screening (within hours) in microwell arrays with volumes down to 2-3 microliters. In addition, we demonstrate the ability of the platform to screen cellulases in microscale volumes at elevated temperatures (up to 95°C) with minimal evaporation. In this study, we utilize our platform to express and screen a panel of beta-glucosidases and cellobiohydrolases isolated from thermophilic microorganisms. Our approach is well suited for a first pass screening of large libraries to identify cellulases with desired properties which can subsequently be produced on a large scale for detailed structural and functional characterization.

See more of this Session: Advances in Proteomics: New Technologies
See more of this Group/Topical: Topical 3: 2010 Annual Meeting of the American Electrophoresis Society (AES)

(198c) A Novel Approach for Modified Protein Identification

Richard Baliban[1], Peter A. DiMaggio Jr.[1], Benjamin A. Garcia[2] and Christodoulos A. Floudas[3], (1) Chemical Engineering, Princeton University, Princeton, NJ, (2)Department of Molecular Biology, Princeton University, Princeton, NJ, (3)Department of Chemical and Biological Engineering, Princeton University, Princeton, NJ

Over the past several years, various tools have been developed for proteomics analysis at the peptide level using tandem mass spectrometry. However, an analysis of a sample data set will not be complete without identification of the comprehensive protein list that is present in the sample. Further, it is imperative that each of these proteins be labeled with all corresponding post-translational modifications (PTMs) to further facilitate the analysis of the dynamic proteome. Though a variety of instruments are available for this analysis, the most accurate measurements come from using high-resolution data from QTOF, LTQ-FT, or Orbitrap instruments. The high accuracy of these instruments can yield better peptide identification, but it is still imperative to ensure that the false positives are minimized at the protein level. A common methodology is to group proteins based on homology, so we have look to recent developments in bi-clustering techniques [1] to provide a useful way to reduce the false positives by grouping together homologous proteins.

To fully analyze a LC-MS/MS data set, we have developed PILOT_PROTEIN, a novel method designed to extract a comprehensive protein list and subsequentaly analyze the uncharacterized MS/MS using an untargeted post-translational modification search. PILOT_PROTEIN [2] will initially use the results of the hybrid peptide identification algorithm PILOT_SEQUEL [3] to generate a list of proteins present in a data sample. The PILOT_SEQUEL algorithm will output a rank-ordered list of the top database peptide sequences and their corresponding proteins [3]. PILOT_PROTEIN will then construct the complete list of possible proteins and assign each protein all of the appropriate peptides from the data set [2]. A protein score is calculated based on the individual scores of the peptides. Due to the large number of proteins on the comprehensive list, a thorough false positive elimination algorithm is employed. Initially, all protein scores are analyzed to see if they pass a minimum threshold. If the score is below the threshold, the protein is eliminated from the possible list. The second stage of the false positive filtering utilizes the clustering algorithm OREO [1] is to group proteins of similar sequence based on their Smith-Waterman alignment score. Only the protein that scores the highest for each cluster is considered as a valid protein while the other proteins are listed as homologues. Once the filtered protein list has been constructed, PILOT_PROTEIN, will analyze all MS/MS that were not annotated spectra using the PILOT_PTM algorithm. Using a universal list of modifications, PILOT_PTM will assign to a template sequence the set of modifications that best explains the experimental data. Derivation of the template sequences will begin with sequence tag generation followed by extraction from the protein list.

To verify the protein prediction capability of the method, PILOT_PROTEIN was initially tested on several high-resolution data sets including (a) 37 spectra fragmented via QTOF and (b) 401 spectra fragmented via Orbitrap [3]. Large-scale LC-MS/MS data sets were then used from the Standard Protein Mix Database which includes (c) 10 complete LC-MS/MS from Orbitrap, (d) 55 complete LC-MS/MS from LTQ-FT, and (e) 30 LC-MS/MS from QTOF [4]. Each data set contains a known set of proteins and sets (c), (d), and (e) contain several low-abundance contaminants. The results were quantified by noting how many of the known proteins are found, how many contaminants were found, and the number of false positives (any additional proteins). PILOT_PROTEIN is able to extract a large number of the possible proteins while significantly reducing the number of false positives reported. The ability of PILOT_PROTEIN was then benchmarked on all five data sets using several known algorithms including InsPecT [5], X!Tandem [6], ProteinProspector [7], and VEMS [8]. PILOT_PROTEIN was able to report a higher number of proteins on average than any of the four competing algorithms and maintains a smaller list of false positives for each data set. The sequence tag generation, candidate peptide extraction, and untargeted modification search were finally tested on 50 LC-MS/MS total chromatin extraction data samples [2].

[1] P. A. DiMaggio Jr., S. R. McAllister, X. L. Feng, J. D. Rabinowitz, H. A. Rabitz, and C. A Floudas. BiClustering bia optimal re-ordering of data matrices in systems biology: rigorous methods and comparative studies. BMC Bioinformatics, 9(1):458, 2008.

[2] R. C. Baliban, P. A. DiMaggio Jr., M. D. Plazas-Mayorca, N. L. Young, B. A Garcia, and C. A. Floudas. A Novel Method for Untargeted Post-Translational Modification Identification Using Integer Linear Optimization and Tandem Mass Spectrometry. Mol. Cell Proteomics, 9:764-779, 2010.

[3] P. A. DiMaggio Jr., B. Lu, J. R. Yates III, and C. A. Floudas. A Hybrid Method for Peptide Identification Using Integer Linear Optimization, Local Database Search, and Quadrupole Time-of-Flight or OrbiTrap Tandem Mass Spectrometry. J. Proteome Res., 7(4):1584–1593, 2008.

[4] J. Klimer, J. S. Eddes, L. Hohmann, J. Jackson, A. Peterson, S. Letarte, P R. Gafken, J. E. Katz, P. Mallick, H. Lee, A. Schmidt, R. Ossola, J. K. Eng, R. Aebersold, and D. B. Martin. The Standard Protein Mix Database: A Diverse Data Set To Assist in the Production of Improved Peptide and Protein Identification Software Tools. J. Proteome Res., 7(1):96–103, 2008.

[5] S. Tanner, H. Shu, A. Frank, L. C.Wang, E. Zandi, M. Mumby, P. A. Pevzner, and V. Bafna. InsPecT: Identification of Posttranslationally Modified Peptides from Tandem Mass Spectrometry. Anal. Chem., 77(14):4626–4639, 2005.

[6] R. Craig and R. C. Beavis. TANDEM: matching proteins with tandem mass spectra. Bioinformatics, 20(9):1466–1467, 2004.

[7] K. R. Clauser, P. R. Baker, and A. L. Burlingame. Role of accurate mass measurement (+/- 10 ppm) in protein identification strategies employing MS or MS/MS and database searching. Anal. Chem., 71 (14):2871–2882, 1999.

[8] R. Matthiesen, M. Lundsgaard, K. G. Welinder, and G. Bauw. Interpreting peptide mass spectra by VEMS. Bioinformatics, 19(6):792–793, 2003.

See more of this Session: Advances in Proteomics: New Technologies
See more of this Group/Topical: Topical 3: 2010 Annual Meeting of the American Electrophoresis Society (AES)

(198d) Generation of Phosphorylated Protein Standards for 2D Gel Western Blotting

Nancy Kendrick, Matt Hoelter and Jon Johansen, Kendrick Labs Inc, Madison, WI

Differentiation pathways in higher organisms are regulated by key protein kinases that phosphorylate and thereby modify the functional state of many substrate proteins. While protein arrays exist for measurement of the kinases, identification of the multiple substrate proteins is difficult. One promising approach is 2D gel Western blotting using antibodies against phosphotyrosine, phosphoserine and phosphothreonine. The phosphorylated proteins that light up can subsequently be identified by mass spectrometry. However, while many anti-phosphoprotein antibodies are commercially available they vary widely in specificity. Purified phosphorylated protein standards, not commercially available except for casein, would be useful in screening antibodies and as positive controls for 2D Western blotting, immunoprecipitation and mass spectrometry. In this presentation we described a method for generating purified phosphorylated proteins in vitro using recombinant kinases and substrate proteins in combination with 32P labeling. The labeled proteins are run on 2D gels with Coomassie blue staining and the phosphorylated isoforms unequivocally identified by autoradiography. When the kinase reaction is carried out with unlabeled 32P, 2DE is used to verify that phosphorylation has occurred by matching the cold Coomassie patterns to those from the 32P-labeled sample. Results will be presented for 3 kinase reactions.

See more of this Session: Advances in Proteomics: New Technologies
See more of this Group/Topical: Topical 3: 2010 Annual Meeting of the American Electrophoresis Society (AES)

(198f) Electrophoretic Separation of Amyloid Proteins Via Capillary Electrophoresis

Elizabeth Pryor, Chemical Engineering, University of Arkansas, Fayetteville, AR, Christa Hestekin, Cheg, University of Arkansas, Fayetteville, AR and Melissa Moss, Chemical Engineering, University of South Carolina, Columbia, SC

The complete sequencing of the human genome has opened the door for the next phase of research, proteomics. Proteomics is a rapidly expanding field and has practical applications for studies of disease-related proteins, such as amyloid proteins. Amyloid proteins, such as insulin and amyloid β (Aβ), are capable of undergoing self-assembly to form oligomeric species and eventually insoluble aggregates. The formation of these species has been linked to the development of diseases including Diabetes and Alzheimer's Disease (AD). A number of technical challenges exist for the quantitative analysis of the different sizes of amyloid proteins which are present during the early stages of aggregation, therefore requiring new technologies to be developed for enhanced protein separations. Microchannel electrophoretic techniques have emerged as powerful tools for the quantitative analysis of proteins. In addition, diagnostic applications of microchannel techniques for the early detection of Aβ protein may offer a means to diagnose and/or prevent AD. This work utilizes capillary electrophoresis (CE) for studies of the early stages of insulin and Aβ aggregation. Previously, we investigated the ability of CE to monitor insulin oligomerization and aggregation over time and worked to develop a model for calculating species sizes. Further, we determined that FITC-labeled insulin was unable to incorporate into the growing aggregation species. Here we extend our previous work to other amyloid proteins, Aβ 1-40 and 1-42.

See more of this Session: Advances in Proteomics: New Technologies
See more of this Group/Topical: Topical 3: 2010 Annual Meeting of the American Electrophoresis Society (AES)

(258a) Pneumatically Controlled 32 Channel Scalable Disposable Microfluidic Sample Handling Device with Integrated Metering, Mixing, and Demultiplexing

Greg A. Liddiard, Department of Electrical and Computer Engineering, University of Utah, Salt Lake City, UT and Bruce K. Gale, Department of Mechanical Engineering, University of Utah, Salt Lake City, UT

As the popularity of integrated LOC systems grows the need for sterilization tolerant or disposable components will follow. Many LOC systems must be carefully sterilized and regenerated. These procedures can often be costly, time consuming, and in some cases require manual intervention. Also, the additional resources required to implement such procedures, i.e. pumps, valves, solution storage, etc., can be undesirable. They can also contribute to the possibility of inaccurate results if not carried out properly. Additionally the device material may have a limited number of sterilization cycles before its efficacy is compromised. Geometry can also contribute to the difficulty associated with sterilization. In many LOC systems this challenge is addressed by simply disposing of the entire device after a single test thus eliminating any possibility of cross-contamination. While the method is scientifically effective many times it is not cost effective. This paper presents the design, fabrication, and testing of a device capable of multiple disposable tests on a single soft lithography fabricated device. The device combines the cost effectiveness of sterilization and reuse of large-area resources as well as the additional robustness of disposable small-area components. The device is constructed exclusively of polydimethylsiloxane or PDMS. Portions of the device were cast using soft lithography and xurography while other components, such as membranes, were created by spinning uncured PDMS onto polystyrene substrates. The construction is a sandwich type. The device consists of six functionally important layers. The first two layers implement the metering system in which volumetrically identical samples can be created. This subcomponent is reusable and serves as the first step before samples are directed to one of the 32 pathways. The third and fourth layers implement the demultiplexing of the sample as well as a separate demultiplexing of an additional reagent. The fifth and final layers implement the disposable reaction chambers and route the sample to an interface for use in further manipulation. The mux assemblies function identically. The control lines actuate both devices simultaneously. The addressing scheme implemented for pathway selection in the system is a based on a simple binary solution commonly used in electronic multiplexors. The pneumatic solution however, requires twice the number of control lines. This is due to the number of valves necessary to simply implement the demux/mux and the single pole single throw nature of the pneumatic valve. The relationship is as follows; where n is equal to the number of valves and f is equal to the number of filters $f=2^{(n/2)}$. This relationship becomes exponentially beneficial as the number of individual filters is increased. Note that n must be an even nonzero value. This system has shown to integrate both disposable and reusable components and utilize the unique benefits of each. The system has also been designed such that additional reusable and disposable components can be added in series or parallel by re-implementing one of the already existing layers.

See more of this Session: Biomedical Diagnostics I
See more of this Group/Topical: Topical 3: 2010 Annual Meeting of the American Electrophoresis Society (AES)

(258b) A Fully Automated Microfluidic Platform for Nucleic Acid Extraction

Michael A. Johnson[1], Jungkyu Kim[2], Bruce K. Gale[1] and Angela Williams[3], (1)Mechanical Engineering, University of Utah, Salt Lake City, UT, (2)Bioengineering, University of Utah, Salt Lake City, UT, (3)Integrated Explorations, Guelph, ON, Canada

Complete lab-on-a-chip sample-in, answer-out systems for nucleic acid analysis would have a wide range of applications from clinical diagnostics to environmental monitoring to genetics research. A key component in such a device is the ability to extract and purify the genetic material (DNA or RNA) in a timely, easy, reliable fashion. We present here a fully automated system capable of extracting RNA at levels comparable with manual commercial extraction methods. Purified RNA samples are delivered in less than 15 minutes from sample input. The system is highly portable, requiring only power input, a connection to a controlling computer, and an optional pressure supply.

The automated system contains a polydimethylsiloxane (PDMS) microfluidic platform complete with on-chip valves, reservoir pumps, and a disposable extraction filter. Fabrication of the PDMS chip involves a modified xurographic technique for multi-level structures. Molds were created using a scotchcal vinyl tape (3M) cut on a sign plotter (Graphtec), with higher structures being cut from acrylic on a carbon dioxide laser and positioned over the tape with an adhesive. The control and operational fluidic layers are separated by a flexible silicone membrane, and the layers are bonded using a novel masked corona discharge method.

Samples and chemicals are mixed and manipulated on the PDMS chip to isolate RNA molecules via solid phase extraction on a glass fiber filter. The system is controlled by a LabView program operating a sequence of solenoid valves, which in turn actuate the on-chip valves and reservoir pumps. The system is thus programmable and adaptable to a variety of sample types.

The device was used to demonstrate successful RNA recovery at yields approaching results attained by a Qiagen commercial spin kit . Known concentrations of RNA from an RNA standard stock (Invitrogen) were used as the sample. RNA extraction results are determined using a fluorescent Ribogreen RNA quantification kit obtained from Invitrogen. Fluorescence data is obtained using a 96-well plate reader. Binding of the nucleic acid to the PDMS surfaces was minimized using various coatings and solution stabilizers. Results from two such coating tests are presented, and extraction yields are compared to the commercial system. RNA recovery yields from a sodium poly-phosphate (NaPP) coated system (35.7 ng) approached those from the commercial spin kit (40.2 ng). However, these high-yield products displayed some NaPP contamination that interfered with reverse-transcript polymerase chain reaction (RT-PCR) analysis methods. A polyvinylpyrrolidone (PVP) coating was used to produce RT-PCR compatible outputs at the cost of a slightly reduced extraction yield (15.8 ng).

The robust nature of the system was demonstrated by extracting RNA from raw biological samples (E. coli cells) extracting up to 429 ng from samples containing 70 million bacteria cells. This highly portable, fully automated system shows promise for more widespread, universal sample preparation for RNA analysis.

See more of this Session: Biomedical Diagnostics I
See more of this Group/Topical: Topical 3: 2010 Annual Meeting of the American Electrophoresis Society (AES)

(258c) Advancement of the Microfluidic Mars Organic Analyzer Platform for Detection of Organic Biomarkers: Amines, Amino Acids, Aldehydes, Ketones, Carboyxlic Acids, and PAHs

Thomas N. Chiesl[1], Amanda Stockton[1] and Richard A. Mathies[2], (1)Chemistry, University of California Berkeley, Berkeley, CA, (2)Department of Chemistry, University of California, Berkeley, Berkeley, CA

Previously, our group created the Mars Organic Analyzer (MOA), microchip CE system for amine and amino acid analysis. Here we present recent work that has extended the MOA's capabilities, via new derivatization and separation chemistries, for analysis of aldehydes, ketones, carboxylic acids, and polycyclic aromatic hydrocarbons. The integration of pacific blue labeling of amino acids increases our sensitivity 250-fold to unprecedented sub parts per trillion levels and permits separations via micellar electrokinetic chromatography (MEKC). The resolution obtained by traditional CZE methods is improved, and when combined with MEKC, pseudo-2D separations allow for enhanced compositional analysis and unambiguous peak identification. Advances in analytical protocols and buffer chemistry have improved the robustness of our CE system when analyzing samples with extreme levels of salinity or pH such as Rio Tino sediments or Saline Valley brine samples. These samples represent places where only extremophiles dwell and, prior to this work, pathologically difficult to analyze by standard electrophoretic methods. We have also analyzed other Martian analog samples (e.g. Murchison Meteorite, Atacama Desert), an environmental sample for PAH analysis, and a simulant of an astrochemistry reaction producing complex organics such as amino acids and dipeptides.

Additionally, oxidized forms of carbon, aldehydes, ketones, and carboxylic acids can now be analyzed with the Mars Organic Analyzer (MOA). Fluorescent derivatization is achieved in ~ 15 min by hydrazone formation with Cascade Blue hydrazide to aldehydes and ketones in 30 mM borate pH 5-6. Labeling of carboxylic acids is achieved by addition of EDC, an activating agent, to the aldehyde and ketone reaction. A carbonyl standard consisting of ten aldehydes and ketones found in extraterrestrial matter and environmental samples is successfully separated. Limits of detection for the components of the standard depend on the reactivity of the compound and range from 70 pM for formaldehyde to 2 uM for benzophenone. To explore the utility of this method for analyzing complex samples, analyses of several wine varietals are conducted, identifying 10 aldehydes and ketones ranging from 30 nM to 5 mM.

Since little is known about the organic chemistry of Mars or other extraterrestrial bodies, any comprehensive organic chemical analysis suite should include the capability to detect and quantify these partially oxidized organic molecules with high sensitivity. Our techniques have additional capabilities required for in situ astrobiological analysis, including the absence of organic solvents, highly fluorescent probes yielding high sensitivity to organic biomarkers, portability, and ease of automation. The chemistries and separations are incorporated into microfabricated bioanalysis system for sample preparation and analysis and are amenable to use in biomedical devices for metabolic analysis.

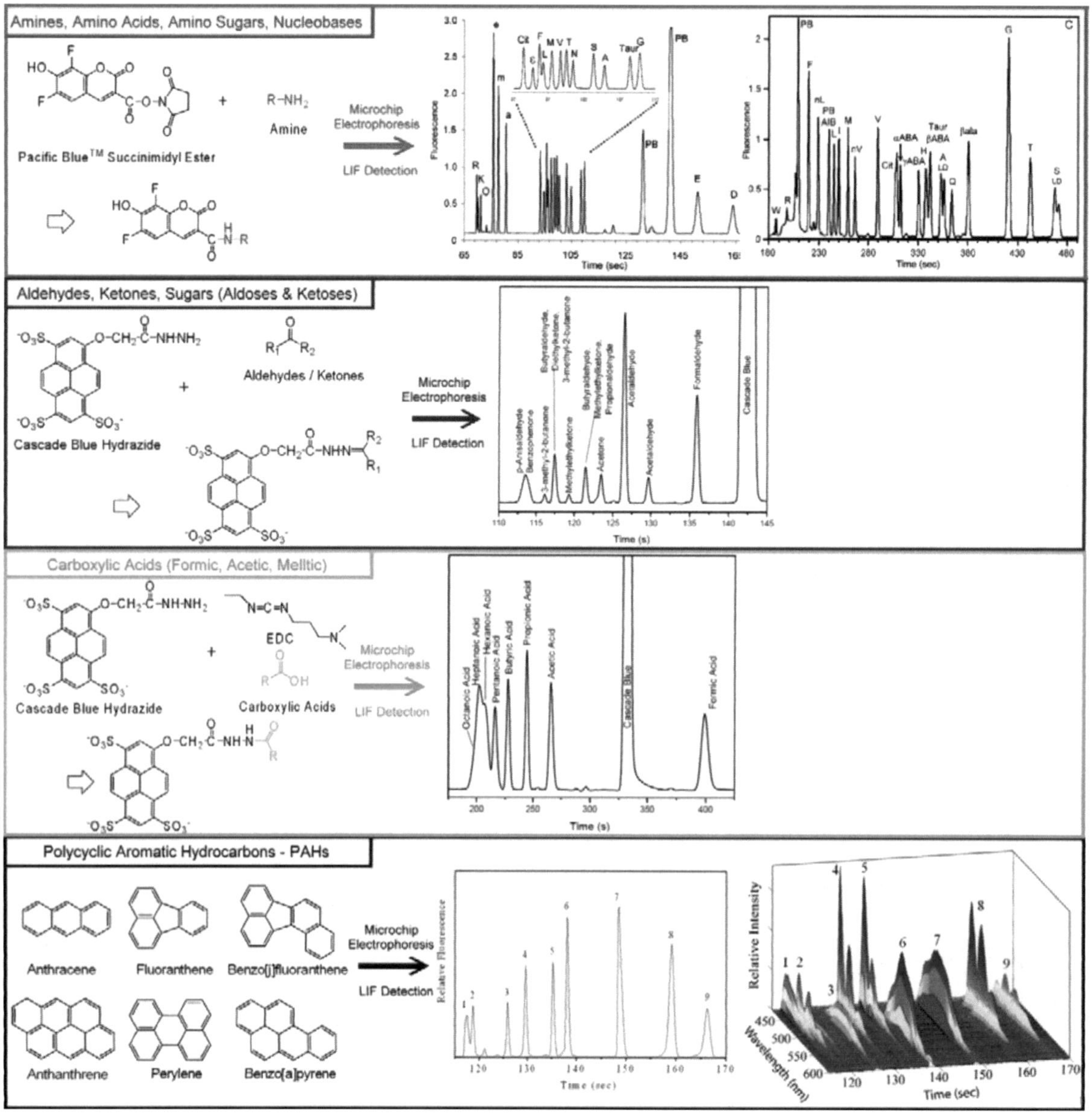

See more of this Session: Biomedical Diagnostics I
See more of this Group/Topical: Topical 3: 2010 Annual Meeting of the American Electrophoresis Society (AES)

(258d) Detection of Rare Variants of Type 2 Diabetes by CE-SSCP

Alice C. Jernigan[1], Hannah Lintag[2] and Christa Hestekin[2], (1)Chemical Engineering, University of Arkansas, Fayetteville, AR, (2)Chemical engineering, University of Arkansas, Fayetteville, AR

Type 2 diabetes is estimated to impact 17.7 million people in the United States (US) and about 171 million people world-wide. The prevalence and seriousness of this disease demonstrates the importance of developing better diagnostic tools and therapeutics. Recent genome-wide association studies of type 2 diabetes have identified at least 15 common variants that increase the risk of type 2 diabetes in populations of European descent. However, these variants individually have small effect, and together account for fewer than 10% of the genetic risk. Therefore, the idea of genetic risk for a disease being conferred by a large number of rarely occurring mutations seems increasingly plausible. These rare variants are characterized by amino acid changes, stop codons, or splice mutations and have large effects. The exploration of the role of rare variants has been hampered by technological challenges. At least several hundred genes could contribute to pathways of insulin action, insulin secretion, or hepatic glucose production, and even exploration of only coding sequences of each gene is a daunting and expensive challenge using current generation sequencing technology. Capillary electrophoresis (CE) single strand conformational polymorphism (SSCP) has the ability to separate DNA not only by size, but by nucleic acid sequence. It can sensitively detect mutations as small as a single base pair alteration. However, CE-SSCP must be optimized and multiplexed in order to be a low cost, high-throughput, and rapid screening method. Currently, we are examining the ability to multiplex CE-SSCP by looking at the simultaneous analysis of multiple DNA fragment sizes (80 – 800 bp) with multiple fluorescent dyes. In order to maintain sensitive mutation detection, three mutation types (substitution, insertion, and deletions) are being studied, along with the ability to resolve them as fragment size increases. Translation to a microfluidic chip is also being explored.

See more of this Session: Biomedical Diagnostics I
See more of this Group/Topical: Topical 3: 2010 Annual Meeting of the American Electrophoresis Society (AES)

(258e) Microfluidic Flow Assays for Diagnosing Bleeding and Thrombotic Disorders

Keith B. Neeves, Department of Chemical Engineering, Colorado School of Mines, Golden, CO, Ryan R. Hansen, Colorado School of Mines, Golden, CO and Jorge A. Di Paola, University of Colorado, Denver

Bleeding and thrombotic disorders are characterized by significant phenotypic variability. Over the last decade important advances have been made in the understanding of the molecular basis of these disorders; however a large portion of the observed variability remains unknown. Current assays for the diagnosis of bleeding disorders have limitations, particularly for the identification of mild to moderate bleeding and the discrimination of common variations in the hemostatic system. Also, quantifying clinical bleeding is extremely difficult. Therefore, a method that shows high sensitivity and specificity is highly desirable.

We have developed a microfluidic flow assay (MFA) to quantify platelet function and coagulation under physiologic flow conditions [1]. The MFA consists of a series of microfluidic channels (250 um X 50 um) in which whole blood is perfused over micropatterned prothrombotic proteins over a range of wall shear rates (Fig 1). The micropatterned proteins include type 1 collagen, the primary adhesive ligand of platelets, and lipidated tissue factor, the initiator of extrinsic pathway of coagulation. We have shown that the focal "injury" in the MFA correlates will with the animal models of thrombosis and can be used to detect dysfunction in platelet function [2,3].

In this talk I will describe the application of the MFA to diagnosing bleeding disorders. We are currently in the middle of a large clinical evaluation of the MFA in collaboration the Hemophilia and Thrombosis Center at the University of Colorado, Denver in which we are screening 500 individuals with bleeding disorders and 1000 controls. Preliminary results with type 1 von Willebrand disease patients show strong correlation between platelet accumulation in the MFA and clinical phenotypes.

[1] K.B. Neeves, S.F. Maloney, K.P. Fong, A.A. Schmaier, M.L. Kahn, L.F. Brass, and S. L. Diamond. Microfluidic focal thrombosis model for measuring murine platelet deposition and stability: PAR4 signaling enhances shear-resistance of platelet aggregates. Journal of Thrombosis and Haemostasis, 6 (2008), 2193-2201.

[2] L. Zhu, T.J. Stalker, K.P. Fong, H. Jiang, A. Tran, I. Crichton, E.K. Lee, K.B. Neeves, S.F. Maloney, H. Kikutani, A. Kumanogoh, E. Pure, S.L. Diamond, and L.F. Brass. Disruption of SEMA4D ameliorates platelet hypersensitivity in dyslipidemia and confers protection against the development of atherosclerosis. Arteriosclerosis, Thrombosis, and Vascular Biology, 29 (2009), 1039-1045.

[3] A.A. Schmaier, Z. Zou, A. Kazlauskas, L. Emert-Sedlak, K.P. Fong, K.B. Neeves, S.F. Maloney, S.L. Diamond, S.P. Kunapuli, J. Ware, L.F. Brass, T.E. Smithgall, K. Saksela, M.L. Kahn. Molecular priming of Lyn by GPVI enables an immune receptor to adopt a hemostatic role. Proceedings of the National Academy of Sciences of the USA, 106 (2009), 21167-21172.

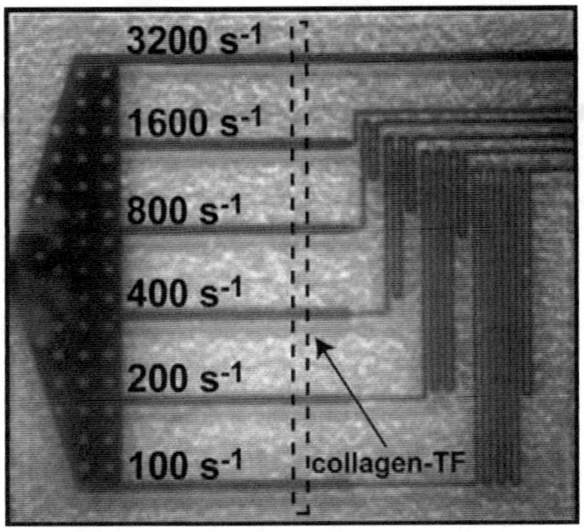

See more of this Session: Biomedical Diagnostics
|
See more of this Group/Topical: Topical 3: 2010 Annual Meeting of the American Electrophoresis Society (AES)

(258f) Electrochemical Biomolecule Sensing by Carbon Nanotubes: Quantification by Warburg Impedance

Satyajyoti Senapati, Sagnik Basuray and Hsueh - Chia Chang, Department of Chemical and Biomolecular Engineering, University of Notre Dame, Notre Dame, IN

Development of rapid and portable detection devices for point-of-care application is an important aspect of modern medical diagnostic industry for effective diagnosis of diseases that result in deaths of millions each year in developing countries. Conventional laboratory based microarray technology is relatively slow, employs multistep procedures, and uses bulky and expensive fluorescent detection units operated by trained staff. The cumbersome equipment requirements restrict the usage of such systems to the laboratory settings. Recently, our group developed a electrochemical DNA detection platform with picomolar sensitivity and a 20 minute detection time using equipment that could be easily turned into a hand held, portable device operated by workers with minimal instruction. The system utilizes an open-flow carbon nanotube platform for DNA detection using impedance based sensing of hybridization between target sequences and short oligonucleotide probes. The open flow allows rapid concentration of the target molecules and shear-enhanced specificity. Interdigitated electrodes are used to trap multiwalled carbon nanotubes functionalized with oligonucleotides via amide bonds. Hybridization of the target sequences with the probes changes the electrochemical impedance spectroscopy. In particular, the Warburg element of the electrochemical sensor shows a fingerprint signature, which allows quantification of target molecules captured by the probe-functionalized sensor. The electrochemical spectrum in a batch micro-fluidic chip is easily explained through the Randles circuit, but further circuit elements are needed in the open flow microfluidic chip to account for high-Peclet number mass transfer and target molecule conformation changes due to shear. This accurate analysis of the circuit in the electrochemical sensor gives a quantitative tool to calibrate the DNA concentration in the sample. Here we also expand this platform's capabilities for the detection of RNA and also antibody/antigen. By focusing on RNA, which naturally occurs with a high copy number, PCR or RT PCR steps can be avoided. RNA-DNA duplexes are more stable than their DNA-DNA relatives, so hybridization rates could be enhanced. Further, with antibody/antigen scheme, significant impedance change recorded on addition of antigen to the antibody functionalized CNT, indicates successful binding of antigens. The impedance changes resulting from binding of a multitude of biomolecules to CNTs shows the promise of the platform for a host of point of care diagnostic devices.

See more of this Session: Biomedical Diagnostics I
See more of this Group/Topical: Topical 3: 2010 Annual Meeting of the American Electrophoresis Society (AES)

(258g) Response of Human Erythrocytes to 10 Khz to 50 MHz Alternating Current Dielectrophoretic Stimulus

Kaela M. Leonard and Adrienne Minerick, Department of Chemical Engineering, Michigan Technological University, Houghton, MI

The use of human blood in medical diagnostics has proven to be an efficient means of obtaining a variety of information about the patient without being too invasive. According to The Merck Manual of Home Health Information, over 65 medical tests can be performed on human blood samples that provide health care workers with information such as specific enzyme, protein and vitamin levels as well as information about the composition of the patient's blood and the oxygen and carbon dioxide saturation levels (Merck Manual of Home Health Information, http://www.merck.com/mmhe/appendixes/ap2/ap2b.html). This information is extremely useful when making a diagnosis and in some cases can only be obtained by way of blood because saliva contains just over 20% of the proteins found in blood. Medical microdevices are poised to obtain quantitative diagnostic information from human blood with improvements in analysis time, reproducibility, and portability over tradition laboratory scale medical diagnostics. The ability to deduce membrane molecular expression would transform medical phenotyping and in the context of human blood would aid in determining a key medical factor - ABO blood type. Previous research by our group showed that O+ could be distinguished from A+, B+ and AB+ at a 95% confidence level and all four blood types could be distinguished at a 56% confidence level when subjected to a 1MHz alternating current field of field density 0.025Vpp/micron. This work is focused on qualitatively and quantitatively determining the differences between blood types when subjected to alternating current dielectrophoretic forces of frequencies between 10 kHz and 50 MHz. Blood was obtained from consenting donors of known blood type and centrifuged down to whole erythrocytes. The erythrocytes were then resuspended in an isotonically matched phosphate buffer saline before being injected into a custom fabricated microdevice. The microdevice was then hooked up to a signal generator that provided the specified frequency, 0.06Vpp/micron AC signal to the perpendicularly positioned micropatterned electrodes. A Zeiss Axiovert Inverted Light microscope was used to take still frame images of the erythrocytes in the microdevice every 15 seconds for 15 minutes. At the end of each experiment the still frame images were run through a customized Zeiss Automeasurement Program to identify each erythrocyte in the field of view as well as obtain quantitative information about the size, shape and placement of the cell. The impact of this research is a potentially easier way to determine the blood type of an unknown sample and later determine any disease markers.

See more of this Session: Biomedical Diagnostics I
See more of this Group/Topical: Topical 3: 2010 Annual Meeting of the American Electrophoresis Society (AES)

(258h) Insulator Based Dielectrophoresis: Dependence of Erythrocyte ABO Antigens

Soumya K. Srivastava, Voiland School of Chemical Engineering and Bioengineering, Washington State University, Pullman, WA, Adrienne Minerick, Department of Chemical Engineering, Michigan Technological University, Houghton, MI and Blanca H. Lapizco-Encinas, Microscale Bioseparations Laboratory, CINVESTAV-Monterrey, Apodaca, Mexico

This research describes a micro-technology application quantifying the dependence of erythrocyte responses by ABO blood type via Direct Current insulator Dielectrophoresis (DC-iDEP). The DC-iDEP device utilized a 330 micron wide rectangular insulating obstacle in a 200 micron channel to create spatial non-uniformities in field density. Just past the insulating obstacle the main channel bifurcates into four outlet channels each 50 micron wide to achieve sorting based on individual erythrocyte deflection from the insulating obstacle. The device and insulating obstacle were fabricated with PDMS following a traditional soft-photolithography process. The PDMS device was sealed to a glass slide by plasma oxidation. The DC-iDEP particle flow behaviors were investigated with all 8 blood types (A+, A-, B+, B-, AB+, AB-, O+, and O-) in the human ABO blood typing system. Different conductivity buffers ranging from 10 to 900 mS/cm and electric fields from 17.12 to 68.5 V/cm were applied to investigate separation dependencies. An SVM microscope from LabSmith was used to capture video at 30 fps for 5 min. The data analysis was conducted with custom built software from Sandia Labs which takes image intensity profiles at locations prior to and in all four outlet channels. The resulting intensity profiles identify particle flow events in the channel at each position as a function of time. Erythrocyte concentration factors were estimated from the total intensities exhibited by a composite sum of cells in the 5 min experimental run. This lab-on-a-chip technology application could be applied to emergency situations, natural calamities, accidents, etc. for blood typing in an accurate, portable, and minimal error free way.

See more of this Session: Biomedical Diagnostics I
See more of this Group/Topical: Topical 3: 2010 Annual Meeting of the American Electrophoresis Society (AES)

(314a) A Programmable Microfluidic System for Selective RNA or DNA Extraction From Various Raw Biological Samples

Michael A. Johnson[1], Jungkyu Kim[2], Bruce K. Gale[1] and Angela Williams[3], (1)Mechanical Engineering, University of Utah, Salt Lake City, UT, (2)Bioengineering, University of Utah, Salt Lake City, UT, (3)Integrated Explorations, Guelph, ON, Canada

While the number and type of microfluidic analysis techniques has grown in recent years, the practical use of lab-on-a-chip systems to perform wide-ranging laboratory-type tests would be greatly benefited by the development of a sample preparation device that purifies genetic material from raw biological samples. The varied conditions of the sources of such material and the stringent requirements of analysis techniques force system-specific sample preparation operations, which often require time consuming and expensive manual handling. We present a system that promises to be a step towards universal sample preparation in its ability to take samples from various sources and selectively extract and purify nucleic acids (DNA or RNA).

The entire extraction system provides a compact, portable control platform with minimal inputs (electrical power, a usb connector, and an optional pressure supply). The polydimethylsiloxane (PDMS) microfluidic chip that performs the main fluid handling tasks of the system is fabricated in three layers: the fluidic channel layer, the functional flexible membrane layer, and the pneumatic control layer. The PDMS chip contains on-chip valves, channels, reservoirs, and pumps that are all pneumatically controlled via electrically actuated solenoid valves. A LabView computer program operates the solenoid valves in sequence to perform a variety of tasks such as sample input, lysis, incubation, mixing, extraction, cleaning, etc. The PDMS chip also has input ports for the reagents required to extract nucleic acid from a variety of samples.

The nucleic acid is purified on the chip using standard solid phase extraction on a disposable glass fiber filter. A cleaning/disinfectant step has been included to allow for multiple tests on a single PDMS chip, but the PDMS chip is also manifold-mounted for easy replacement for sensitive tests where disposable units are desired.

The extraction process operates by pumping the sample and various buffers and solutions through integrated pipetting/mixing reservoirs and past the extraction filter. Altering the amounts, types, and sequences of reagents used is easily accomplished by adjusting the control program, which allows selective adjustment of the protocol to handle different sample types and control the output as DNA, RNA or both. The ability of the system to handle different sample inputs is demonstrated by extracting DNA from both a stock DNA solution and whole blood, and by extracting RNA form a both a stock RNA solution and living E. coli cells with 429 ng of RNA being extracted from 70 million cells.

This system has proven capable of handling a variety of sample preparation tasks in an effort to make automated nucleic acid sample preparation more practical and available. Such work is a step towards a universal sample preparation unit that can handle many kinds of samples from viruses to tissue samples and provide rapid, meaningful results from a lab-on-a-chip platform.

See more of this Session: Biomedical Diagnostics II
See more of this Group/Topical: Topical 3: 2010 Annual Meeting of the American Electrophoresis Society (AES)

(314b) One-Step Extraction of Subcellular Proteins From Eukaryotic Cells

Yihong Zhan[1], Victoria A. Martin[2], Robert L. Geahlen[2] and Chang Lu[3], (1)Agricultural and Biological Engineering Department, Purdue University, West Lafayette, IN, (2)Medicinal Chemistry and Molecular Pharmacology, Purdue University, West Lafayette, IN, (3)Chemical Engineering, Virginia Tech, Blacksburg, VA

Conventional biochemical analysis mainly focuses on the expression level of cellular proteins from entire cells. However, it has been increasingly acknowledged that the subcellular location of proteins often carries important information. Analysis of subcellular proteins conventionally requires subcellular fractionation which involves two steps: cell lysis to release proteins and high-speed centrifugation to separate the homogenate. Such approach requires bulky and expensive equipment and is not compatible with processing scarce cell samples of limited volume. In this study, we apply microfluidic flow-through electroporation to breach cell membranes and extract cytosolic proteins selectively in a single step. We demonstrate that this approach allows monitoring the translocation of the transcription factor NF-êB from the cytosol to the nucleus without the need of subcellular fractionation. Our technique is compatible with the processing of samples of various sizes and provides a simple and universal tool for bioanalytical analysis and spatial proteomics.

See more of this Session: Biomedical Diagnostics II
See more of this Group/Topical: Topical 3: 2010 Annual Meeting of the American Electrophoresis Society (AES)

(314c) Multichannel Mars Organic Analyzer (McMOA): Microfluidic Networks for the Automated in Situ Microchip Electrophoretic Analysis of Organic Biomarkers

Thomas N. Chiesl, Merwan Benhabib, Amanda Stockton and Richard A. Mathies, Chemistry, University of California, Berkeley, Berkeley, CA

The goal of this work is to achieve autonomous routing and analysis of samples on redundant CE microchannels with negligible sample carryover. The development of microfluidic architectures that integrate automated metering and mixing of fluids for performing on chip reactions and serial dilutions is also discussed. We have successfully built and operated a new prototype instrument, the Multichannel Mars Organic Analyzer (McMOA), that tests key functional aspects required for in situ planetary exploration for organic compounds. The instrument uses a four-layer microchip containing eight CE analysis systems integrated with a microfluidic network for fluidic processing and includes a linear scanning optical excitation system, a charged-coupled device (CCD) fluorescence spectrometer detection system and a temperature controlled chip platform. We have demonstrated serial autonomous amino acid and PAH assays on the same chip. The same device and instrumentation is also capable of performing CE assays of organic sugars, aldehydes, ketones, and carboxylic acids. The CCD spectrometer was used to identify the unique fluorescence spectra of nine components in a PAH standard and then applied to the analysis of a sediment sample from Lake Erie. These results demonstrate the readiness of lab-on-a-chip CE electrophoresis technology to autonomously perform life detection experiments. While the microfluidics and instrumentation discussed here are oriented towards in situ planetary analysis, the engineering principles and results coincide with the development of successful portable rapid-analysis biomedical and environmental diagnostics.

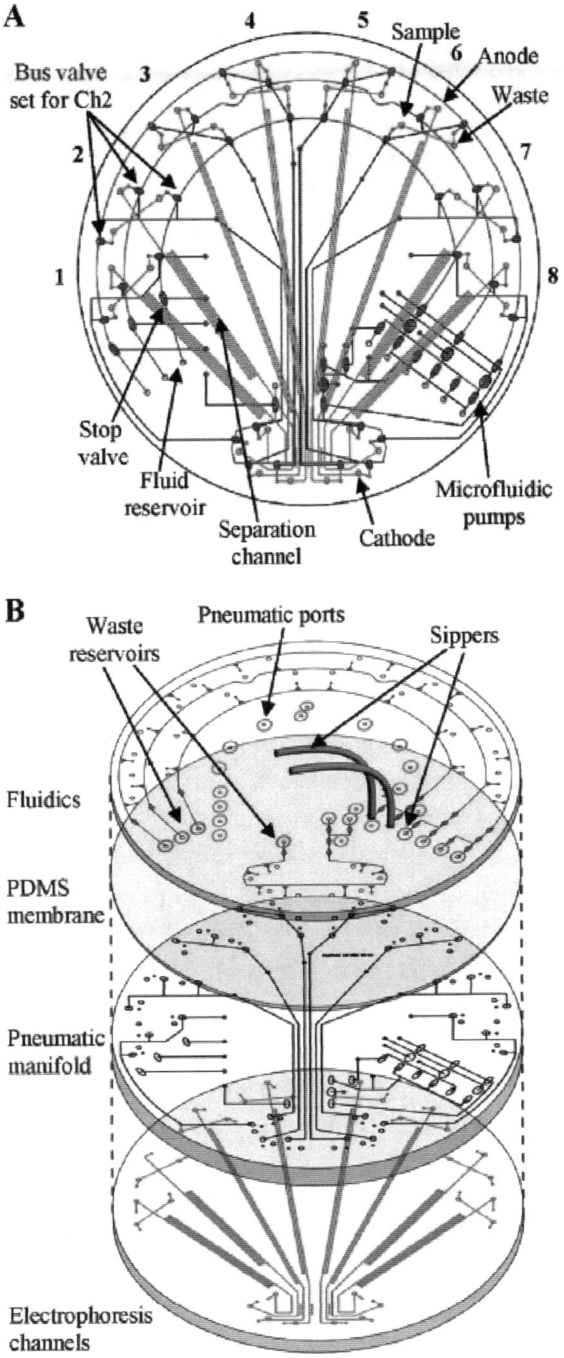

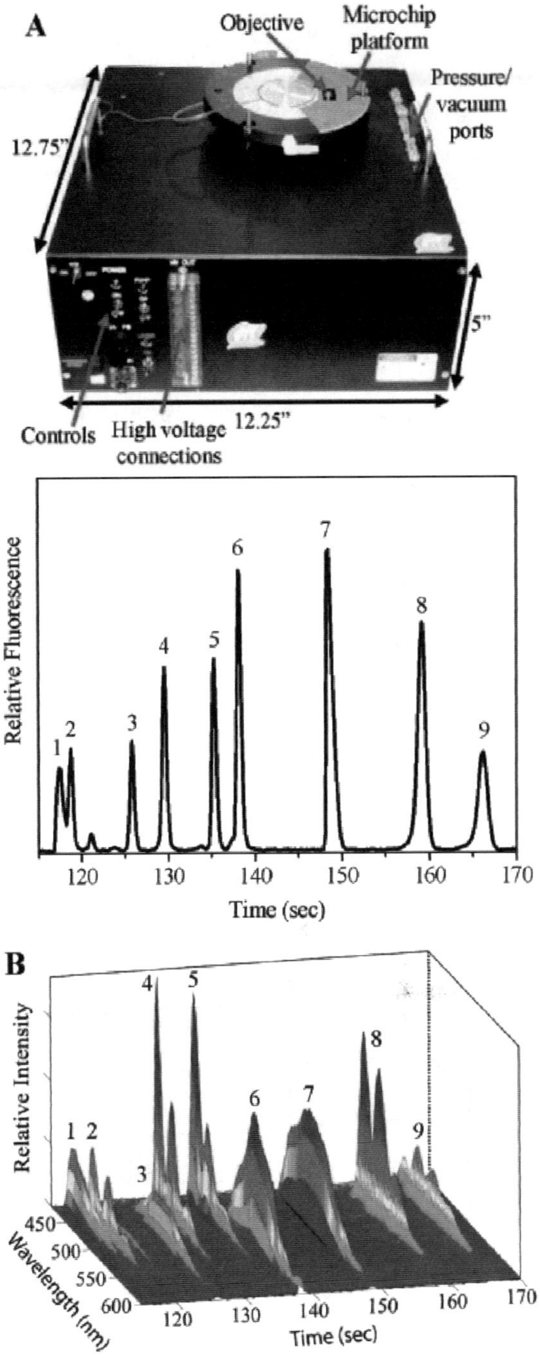

See more of this Session: Biomedical Diagnostics II
See more of this Group/Topical: Topical 3: 2010 Annual Meeting of the American Electrophoresis Society (AES)

(314d) Pneumatically Driven 16 Channel Disposable Nucleic Acid Filter Device with Integrated Demultiplexing and Multiplexing

Greg A. Liddiard, Department of Electrical and Computer Engineering, University of Utah, Salt Lake City, UT, Erik Liddiard, Department of Bioengineering, University of Utah, Salt Lake City, UT and Bruce K. Gale, Department of Mechanical Engineering, University of Utah, Salt Lake City, UT

As the popularity of filter based nucleic acid extraction systems grows the need for regenerable or disposable filters and filter assemblies will follow. A unique advantage of filter based nucleic extraction systems is their ability to fully integrate and implement major extraction system components. Unlike their centrifugation centric counterparts the integrated soft lithography solutions don't require complex, power-hungry, poorly scalable centrifuges to separate NAs from the sample material. Instead, the solution is passed through a filter which provides a binding surface whereby NAs can be temporarily attached by means of a salt bridge. These NAs are later released in the presence of an elution buffer. This method, along with other solid state extraction schemes, has demonstrated extraction efficiencies comparable to those attained with a conventional centrifugation process(ref). A challenge presented by this type of system is the regeneration of the filter material. If proper elution and sterilization is not performed cross-contamination is almost guaranteed. This can be accomplished by means of sterilization assays and regeneration protocols. However, the additional resources required to implement such procedures, i.e. pumps, valves, solution storage, etc., can be undesirable. Additionally the filter material may have a limited number of regeneration cycles before its efficacy is compromised. This paper presents a novel solution that can act as a stand-alone device or can be integrated into a larger system. We propose a disposable or limited use filter ring containing 16 individual filters each addressable by means of pneumatically driven microvalves. The term "ring" is used to describe the disposable filter material portion because the geometry of the reusable mux/demux portion requires that the filters be arranged in a circle or ring. The device offers a single input and single output. All demultiplexing and multiplexing of the solution is handled inside the device. By integrating demux/mux capabilities the device can be quickly integrated into any current single filter system. The only requirements for integration are: the necessary hardware connections, a reasonable tolerance for additional dead volume, and the supply of pneumatic selection lines. The device is constructed exclusively of four materials; polydimethylsiloxane or PDMS, silicone membrane, glass microfiber, and poly methyl methacrylate or PMMA. The PMMA is purely structural and does not contact the solution. The construction is a sandwich type. The device consists of individual isolated glass microfiber filters between PDMS slabs. This structure is compressed between the demux and mux subassemblies. The demux and mux assembly are mirror images and function identically. The control lines actuate both devices simultaneously. PMMA is used for even compression of the device by the clamping mechanisms. Also, a PMMA ring is used to direct compression of the device away from the microchannels, so as not to block the flow of solution, and towards the areas of potential leakage. The addressing scheme implemented for filter selection in the system is a based on a simple binary solution commonly used in electronic multiplexors. The pneumatic solution however, requires twice the number of control lines. This is due to the number of valves necessary to simply implement the demux/mux and the single pole single throw nature of the pneumatic valve. The relationship is as follows; where n is equal to the number of valves and f is equal to the number of filters f=2^(n/2). This relationship is very beneficial as the number of individual filters is increased. Note that n must be an even nonzero value. The device has shown to be easily integrated into existing systems providing a quickly disposable, individually addressable, multiple filter solution for solid state NA extraction.

See more of this Session: Biomedical Diagnostics II
See more of this Group/Topical: Topical 3: 2010 Annual Meeting of the American Electrophoresis Society (AES)

(314e) Chip-Based in Situ Hybridization for Identification of Bacteria From the Human Microbiome

Robert J. Meagher, Peng Liu, Yooli K. Light and Anup K. Singh, Sandia National Laboratories, Livermore, CA

The emerging field of metagenomics seeks to assess the genetic diversity of complex mixed populations of bacteria, such as those found at different sites within the human body. A single person's mouth typically harbors up to 100 bacterial species, while surveys of many people have found more than 700 different species, of which ~50% have never been cultivated. In typical metagenomics studies, the cells themselves are destroyed in the process of gathering sequence information, and thus the connection between genotype and phenotype is lost. A great deal of sequence information may be generated, but it is impossible to assign any given sequence to a specific cell.

We seek non-destructive, culture-independent means of gathering sequence information from selected individual cells from mixed populations. As a first step, we have developed a microfluidic device for concentrating and specifically labeling bacteria from a mixed population. Bacteria are electrophoretically concentrated against a photopolymerized membrane element, and then incubated with a specific fluorescent label, which can include antibodies as well as specific or non-specific nucleic acid stains. Unbound stain is washed away, and the labeled bacteria are released from the membrane. The stained cells can then be observed via epifluorescence microscopy, or counted via flow cytometry. We have tested our device with three representative bacteria from the human microbiome: E. coli (gut, Gram-negative), Lactobacillus acidophilus (mouth, Gram-positive), and Streptococcus mutans (mouth, Gram-positive), with results comparable to off-chip labeling techniques.

See more of this Session: Biomedical Diagnostics II
See more of this Group/Topical: Topical 3: 2010 Annual Meeting of the American Electrophoresis Society (AES)

(314f) Integrated Microfluidics for Serotype Identification of Foot and Mouth Disease Virus

Himanshu Jayant Sant, Mechanical Engineering, State of Utah Center of Excellence for Biomedical Microfluidics, University of Utah, Salt Lake City, UT, Scott Sundberg, University of Utah, SLC, UT, Erik Liddiard, Bioengineering, University of Utah, SLC, UT, Michael A. Johnson, Mechanical Engineering, University of Utah, Salt Lake City, UT and Bruce Gale, Department of Mechanical Engineering, University of Utah, Salt Lake City, UT

This work reports an integrated microfluidics-based biosensor for Foot-and-mouth-disease virus (FMDV) detection. FMDV is highly contagious and prevalent in the most economically important animals worldwide [1]. FMDV exists in seven different types of serotypes and early identification of which is important in understanding the disease and its potential spread in the population.

Our laboratory is developing a field-based analysis tool for rapid FMDV serotype identification from clinical samples that range from vesicular fluid, feet or nose epithelium to whole blood. Specifically, this work shows the integration of an automated microfluidic extraction unit with a spinning disk platform for RT-PCR that allows us to analyze the different FMDV serotypes with high sensitivity in about an hour with current analysis protocol.

Figure 1 is a photograph of the microfluidic extraction system in its current format. The top portion consists of a PDMS microfluidic platform complete with on-chip valves (31 pneumatic valves), reservoir pumps (9 chambers that also work as pumps), and a disposable extraction filter. The system is controlled by a LabView program that operates a sequence of solenoid valves to run the RNA extraction protocol, which uses the same chemical sequence as the Qiagen RNeasy Mini spin kit. The initial experiments have been carried out with E. coli bacterial samples while the appropriate testing conditions for FMDV at our facilities are established. Figure 2 shows the comparison of extraction output with spin kit for stock E. coli RNA sample. The automatically extracted RNA elutant is compared to the output of the manual Qiagen RNA spin kit (RNEasy Kit) as the gold standard for extraction efficiency. RNA extraction results are determined using a Ribogreen RNA quantification kit obtained from Invitrogen. Fluorescent data is obtained using a 96-well plate reader. The latest version of our on-chip protocol has approached the efficiency of the commercial spin kit, and we anticipate further improvement and stabilization.

The disk for RT-PCR is based on the use of centrifugal force to move fluid along 7 short spiral channels with individual inlets located at the center to run separate chemistries for different viral serotypes. This device is fabricated from thin film plastics to create an inexpensive disposable, requiring only centrifugation for fluid control. The entire disk is 120 mm in diameter, the size of a CD, and is 375 μm thick (see Figure 3). The channel layer disks were manufactured using the process of xurography [2].

This platform has already been used to amplify and analyze a 172 bp E. Coli RNA product with encouraging results (using one-step Qiagen RT PCR kit). A rapid air thermocycler with proper cycle times and temperatures was used to perform RT-PCR in single run [3]. Work is underway to further optimize the PCR cycle times, extraction protocol and efficiency improvement (both extraction and PCR) for integrated analysis.

Figure Captions:

Figure 1 Picture of an integrated microfluidic extraction chip

Figure 2 Comparison plot showing ng RNA extracted with standard spin kit and microfluidic chip.

Figure 3 Schematic diagram of the RT-PCR disk showing individual serotype loading chambers and lanes.

References:

1. C. Carrillo, E. R. Tulman, G. Delhon, Z. Lu, A. Carreno, A. Vagnozzi, G. F. Kutish, and D. L. Rock, Comparative Genomics of Foot-and-Mouth Disease Virus. Journal of Virology, 79(10): 6487-6805, 2005.

2. Bartholomeusz DA, Boutte RW and Andrade JD. Xurography: rapid prototyping of microstructures using a cutting plotter. JMEMS, 14(6): 1364-1374, 2005.

3. Sundberg SO, Wittwer CT, Gao C and Gale BK. Spinning disk platform for microfluidic digital polymerase chain reaction. Anal Chem, published ASAP online, accepted December 31, 2009.

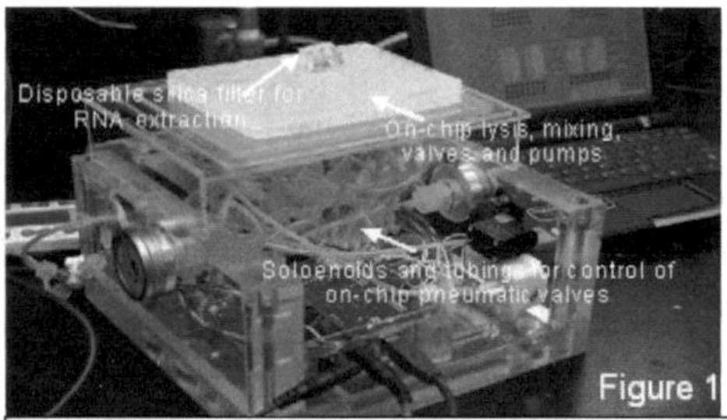

Figure 1

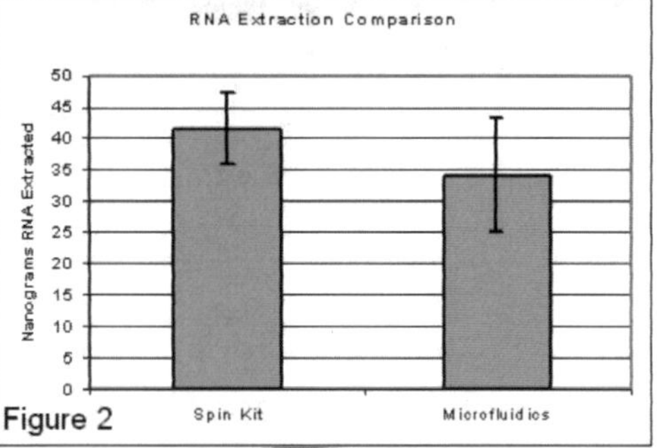

Figure 2

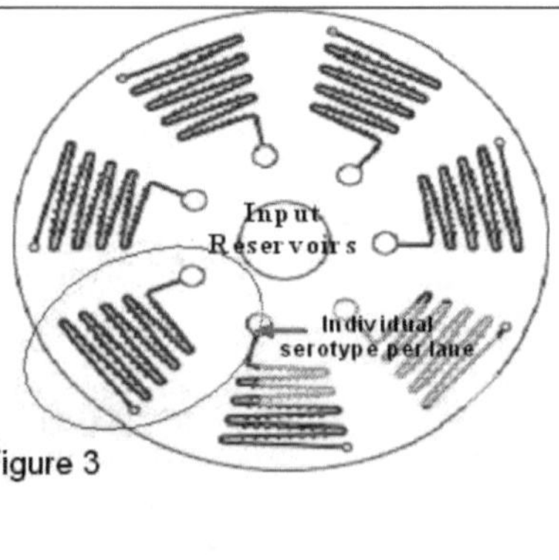

Figure 3

(314g) Asymmetrical Cross-Flow Based Split Thin Cell Fractionation

Venu M. Arremsetty and Bruce K Gale, Mechanical, University of Utah, Salt Lake City, UT

A growing need for the separation of nano particles may be solved by split flow thin cell fractionation (SPLITT). This work introduces a novel cross-flow based SPLITT microsystem that is capable of high speed continuous separations based on size and density of the sample.An Asymmetrical cross flow based SPLITT system has been used to separate 470 nm polystyrene nanoparticles and albumin in less than 15 minutes. Work is still under progress to model the system to identify optimized set of inlet flowrates, total flowrates, cross-flow and other two outlets for high resolution separation. The microsystem consists of a flow unit that consists of two fluidic channels (Of which one channel is made of double sided tape of thickness 50 microns and other one is a thin PDMS layer coated on double sided tape of thickness 100 microns) that are separated by a splitter layer made of stiff Mylar sheet of thickness 125 micron. One side of flow unit utilizes a porous frit that allows for cross-flow in or out of the channel as one wall and glass slide as another wall. Polystyrene particles of size 470 nm and albumin were injected for system characterization. Initial results show that the system is capable of retaining most of the particles and future work will involve optimizing the flowrate combinations for improved and continuous separation of more complex samples.

See more of this Session: Biomedical Diagnostics II
See more of this Group/Topical: Topical 3: 2010 Annual Meeting of the American Electrophoresis Society (AES)

(314h) Measurement of Muscle Force in C. Elegans Worm Using Microfluidics

Erik Liddiard, Department of Bioengineering, University of Utah, SLC, UT, Himanshu Jayant Sant, Mechanical Engineering, State of Utah Center of Excellence for Biomedical Microfluidics, University of Utah, Salt Lake City, UT, Frederic Horndli, Department of Biology, University of Utah, SLC, UT and Bruce Gale, Department of Mechanical Engineering, University of Utah, Salt Lake City, UT

We have designed, fabricated, and tested a microfluidic device which allows biologists to better understand the contraction of muscles used in locomotion of the worm Caenorhabditis elegans (C. elegans). Biologists who have genetically altered worms in ways thought to affect the muscle contraction force need a means for comparing mutant worms to wild type worms. Due to the worms' size, traditional methods for measuring muscle force are not feasible. The worm C. elegans has been studied extensively due to its relative simplicity and straightforward use as a model for genetic manipulation. Results from the study of the C. elegans worm have resulted in two Nobel prizes, and show potential for breakthroughs in diseases such as muscular dystrophy. However, due to the worms' small physical size, 1 mm in length and ~80 µm in diameter, manipulation of the worm can be difficult. Microfluidics, which deals with controlling and transporting fluids through channels of micro meter dimensions, has been utilized in studying and controlling the worm because of its ability to confine the worm within small channels and allow for manipulation. Microfluidic devices are often made from polymers, the most common being Polydimethylsiloxane (PDMS). PDMS devices offer the advantage of maintaining the worms in a nontoxic, oxygenated environment. Microfluidic devices have been constructed to trap the C. elegans worm allowing for imaging and surgery, however these devices have not been used in a sensing application. We have developed a device which immobilizes the worm and is capable of distinguishing differences in muscle contraction force between wild and mutant type worms. Our device consists of three layers: a fluid layer, air layer, and a flexible membrane which separates the fluid and air layers. The flexible membrane is used in measuring the force generated by the muscles of the C. elegans worm which are responsible for locomotion. A thin 30 µm flexible membrane is used to immobilize the worm. Using sensitive air regulators the air pressure on the membrane face opposite that of the worm can be finely increased extending the membrane into a fluid channel which contains a worm. As the membrane extends into the channel it contacts the worm and applies a force on the worm. As the air pressure is increased, the force the membrane exerts on the worm increases until the worm becomes immobilized, at which point the air pressure is recorded. Three strains of C. elegans worms were tested in our device, a wild type strain and two strains with genetic knockouts, UNC-29 and ACR-16, which removed ligand gated ion channels on the muscle responsible for contraction. Immobilization pressures for the mutant worms showed statistical differences when compared to the wild type worm allowing us to rank the worm strains according to muscle contraction force (fig. 1). Utilizing microfluidic techniques we have created a device which uses a flexible membrane and sensitive air regulator to establish a relationship between air pressure and the muscle contraction force of the C elegans worm.

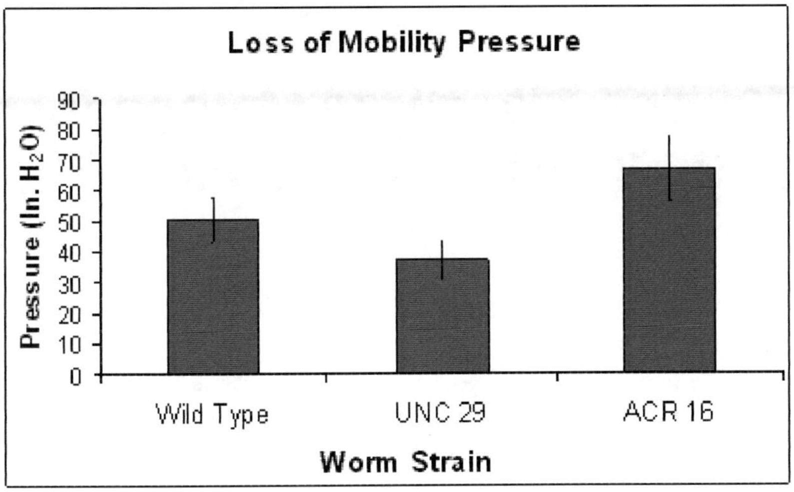

See more of this Session: Biomedical Diagnostics II
See more of this Group/Topical: Topical 3: 2010 Annual Meeting of the American Electrophoresis Society (AES)

(314i) A Masked Corona Discharge Method for Selective Bonding in PDMS for Microfluidic Applications

Michael A. Johnson[1], Erik Liddiard[2] and Bruce K. Gale[1], (1)Mechanical Engineering, University of Utah, Salt Lake City, UT, (2)Bioengineering, University of Utah, SLC, UT

Multi-layer PDMS devices are commonly used for microfluidic applications, and the method for bonding the several layers together has an impact on determining the possible uses for the device. Surface activation bonding methods such as oxygen plasma treatments and corona discharge methods have proven to supply ample adhesion between layers while having minimal adverse effect on the microfluidic device capabilities. This work examines an enhancement to corona discharge bonding methods by using a selective mask to bond only desired portions of the PDMS layers with applications in improving on-chip pneumatic valves and creating zero dead-volume channels and reservoirs.

Corona discharge bonding operates by creating hydrophilic, chemically active functional groups on the PDMS surface that bind to other activated groups when contacting another treated surface. Covering the PDMS layer to be bonded with a masking material during corona discharge treatment blocks areas from exposure to the corona plasma so that only exposed surfaces are activated. Subsequent removal of the mask and contact with another activated PDMS surface is shown to produce selective bonding, with the masked areas remaining unbound. Several masking materials are shown to be effective: Rubylith electrostatic film, Scotchcal Graphtec Instachange tape, mineral oil, and liquid spin-on glass. An acrylic sheet is also cut with a laser to provide a mask manifold to simultaneously mask multiple valve sites. The masked corona bonding method is shown to be effective in bonding PDMS to PDMS, glass, and other silicone materials.

This fabrication method is shown to be an efficient method to create passive check valves or pneumatically actuated valves in a three-layer PDMS device. The valves follow a similar design to other on-chip pneumatic valves, but the valve seat is masked from bonding to prevent it from becoming permanently closed while allowing the surrounding material to be permanently bonded in order for the valve to handle high pressures if needed.

Reservoirs and channels are created in a similar three-layer design with a pneumatic control layer separated from a fluidic layer by a thin membrane. Such features are demonstrated to consistently fill to a desired volume and subsequently empty completely, thus creating effectively zero dead-volume channels and reservoirs without the difficulty of fabricating channels and reservoirs with rounded edges in the vertical direction.

Masked corona discharge bonding in PDMS structures is used to create practical components in microfluidic devices with applications like trapping C. elegans worms and extracting nucleic acid from raw biological samples. This bonding method increases the capabilities of using PDMS as a material for rapid prototyping in microfluidic systems.

See more of this Session: Biomedical Diagnostics II
See more of this Group/Topical: Topical 3: 2010 Annual Meeting of the American Electrophoresis Society (AES)

(372a) Rapid Detection of Bacteria in Blood for the Early Diagnosis of Sepsis

Sachidevi Puttaswamy and Shramik Sengupta, Biological Engineering, University of Missouri, Columbia, MO

Sepsis, (systemic inflammation caused by bacterial infection of blood) is a leading cause of death in the US with a mortality rate of ~ 30%. The clinical symptoms displayed by the patients are subtle and non-specific. At the time clinical symptoms begin to be displayed, the concentration of bacteria present in their blood is often as low as 1-10 CFU/ml. If allowed to progress, the disease leads to multiple organ failure and ultimately death. Multiple studies have shown that the quicker one can detect the presence of bacteria, and start effective antibiotic therapy, the greater is the likelihood of survival for the patient. The current technologies for this purpose are Automated Blood Culture Systems such as the BACTEC, BacT/Alert and VITEK, which take 1-5 days to obtain the result.

Unlike the current culture systems that detect the presence of bacteria in the suspensions indirectly via the effects of bacterial metabolism (change in pH of levels of O_2/CO_2), we use a novel technique that relies on the fact that bacteria can store large amount of charge and thus increase the bulk capacitance of the liquid in which they are suspended. An increase in the number of bacteria leads to an increase in the bulk capacitance of the suspension. Usually changes in the bulk capacitance are masked by the electrochemical interfacial (double layer) capacitance. But by leveraging microfluidic geometry, and taking measurements at multiple frequencies, we are able to accurately monitor changes in the bulk capacitance. Doing so enables us to pick up signatures of bacterial presence in the suspension at concentrations of ~ 1000 cfu/ml (as opposed to > 10^6 cfu/ml for current technologies), and detect the presence of ~10 E coli /ml within 4 hours, as opposed to ~20 hours required by current technologies.

See more of this Session: Poster Session for the American Electrophoresis Society
See more of this Group/Topical: Topical 3: 2010 Annual Meeting of the American Electrophoresis Society (AES)

ELECTROKINETIC DIFFERENTIATION OF MICROORGANISMS IN GLASS MICROCHANNELS

Lucía D. Garza-García[1], *Victor H. Pérez-González*,[1] *Oscar A. Pérez-Sánchez*[1] *and Blanca H. Lapizco-Encinas*[2,*] *Email:* blapizco@cinvestav.mx

[1]*Tecnologico de Monterrey, Monterrey, N.L., Mexico*
[2]*Centro de Investigación y de Estudios Avanzados del IPN Unidad Monterrey, Apodaca, N.L., México*

Abstract

Microscale electrokinetic techniques are becoming increasingly important for the manipulation of microorganisms; potential applications are in the fields of clinical analysis, environmental monitoring, and food safety detection. This study demonstrates the electrokinetic characterization of three different species of microorganisms (*Ankistrodesmus sp.*, *Escherichia coli,* and *Saccharomyces cerevisiae*) by employing micro particle image velocimetry. Microorganisms were introduced in a straight glass microchannel, where direct current electric fields were applied (50-300 V/cm). Experimental results allowed identifying electrokinetic velocity and mobility for each type of microorganism, as well as the electrokinetic differentiation of the three different types of cells when mixed in a sample by obtaining an electropherogram. These results demonstrate that microscale electrokinetic techniques have the potential for rapid microbial characterization and identification.

Introduction

Microanalytical systems are becoming increasingly important for the manipulation of microorganisms, with potential applications for food borne pathogen detection, environmental monitoring, and clinical analysis. This accelerated development is driven by the important advantages microsystems offer, such as low sample and reagent consumption, fast processing and response time, portability, and low cost.[1,2] Electrokinetic microdevices are continuously growing as tools for the separation of microorganisms, since they offer high design flexibility, allowing for manipulation and recognition of different cells characteristics; [3,4] with the attractive advantage that electrokinetic techniques allow for the separation of intact cells.

Dielectrophoresis (DEP) and Electrophoresis (EP) are electrokinetic techniques usually applied for the separation and sorting of microorganisms. Dielectrophoresis (DEP) consists in the movement of particles in suspension when exposed to a non-uniform electric field due to polarization effects.[5,6] This microscale technique is gaining popularity for the manipulation and separation of microorganisms.[3,7] In 1968 Crane and Pohl[8] achieved a high efficiency separation of live and dead yeast cells by using AC electric fields with wire electrodes. In the next twenty years DEP did not developed significantly due to the lack of sophisticated electronic equipments and microelectrode arrays. In the last ten years, there has been a spurt of studies on DEP focused on the manipulation of many microorganisms such as virus,[9] bacteria,[10,11] yeast[12] and parasites.[13]

Electrophoresis can be defined as the characterization, manipulation, and quantification of microorganisms based on their differences in charge/mass ratio.[4,14] In 1999 Armstrong *et al.*[4] was able to achieve the separation of bacterial and yeast cells (*P. fluorescens, E. aerogenes, M. luteus,* and *S. cerevisiae*) with the application of capillary electrophoresis (CE) by coating the capillary walls with polyethylene oxide to minimize electroosmotic flow (EOF). As a result an electropherogram was obtained in less than 15 minutes, where the area of each peak was proportional to the number of cells in the sample. In 2008, Jackowski *et al.*[15] achieved the separation of *P. vulgaris* and *E. coli*, as well as the identification of *E. coli* in infected urine samples and *H. pylori* from stomach mucosa samples, obtaining an electropherogram in less than 20 minutes. Since this study was done with capillary zone electrophoresis (CZE), it was demonstrated that this technique can be used in medical diagnosis and clinical analysis of pathogenic bacteria with high accuracy and fast response time.

The EP velocity is described by:

$$v_{EP} = \mu_{EP} E \quad (1)$$

where v_{EP} is the electrophoretic velocity, μ_{EP} is the electrophoretic mobility, and E is the applied electric field. Electroosmotic flow (EOF) occurs due to double layer effects when a direct current (DC) electric field is applied in a capillary tube or microchannel, generating fluid movement. EOF is commonly used for the transportation of liquids in microdevices; the EOF velocity is defined by:

$$v_{EO} = \mu_{EO} E \quad (2)$$

where v_{EO} is the electroosmotic velocity and μ_{EO} is the electroosmotic mobility. Electrokinetics (EK) considers the movement of a particle and fluid, *i.e.,* it considers the effects of EOF and EP present in the same system, the electrokinetic velocity is defined by:[16]

$$v_{EK} = \mu_{EK} E = (\mu_{EO} + \mu_{EP})E \quad (3)$$

where v_{EK} is the electrokinetic velocity and μ_{EK} is the electrokinetic mobility. EP is an important effect when the particles posses a significant charge to mass ratio. In this study, where microorganisms were used, it was possible to neglect the effect of EP since microbes posses a low superficial charge[17] in comparison to their mass. Therefore, electrokinetic mobility is mainly due to EOF effects:

$$\mu_{EK} \approx \mu_{EO} \quad (4)$$

where electroosmotic mobility is defines as:

$$\mu_{EO} = \frac{\varepsilon_m \zeta}{\eta} \quad (5)$$

This study demonstrates the characterization of electrokinetic velocity and mobility of three different species of microorganisms (bacteria *E. coli*, yeast *S. cerevisiae*, and microalgae *Ankistrodesmus sp.*) in a 10-µm-deep glass microchannel, using micro particle image velocimetry (µPIV). Experiments were performed at different values of direct current electric fields and suspending medium properties (conductivity and pH). Results showed that *E. coli* has a greater electrokinetic velocity, followed by *Ankistrodesmus sp.* and then by *S. cerevisiae*, being the bacteria the smallest cell. Therefore, it was demonstrated that microbe electrokinetic mobility depends strongly on cell size when shallow microchannels are employed. Electrokinetic differentiation was achieved by employing a mixture of the three different species, obtaining an electropherogram in less than 2 minutes, demonstrating the great potential of microscale electrokinetic techniques for fast microbial analysis.

Materials and Methods

Microdevice

The microdevice employed was made from glass with a microchannel 30-mm-long, 1-mm-wide, and 10-µm-deep. Each device consisted of two rectangular glass wafers made of Schott 263D (Howard Glass, Worcester, MA, USA). The microchannel was patterned on the bottom wafer using standard wet-etching techniques and the reservoirs were drilled on the top wafer; and as a final fabrication step the two wafers were thermally bonded.

Suspending medium

The suspending medium used consisted of deionized water with adjusted pH and conductivity. By adding NaOH and HCl, pH values were adjusted to 5, 6, 7, and 8; meanwhile, by adding K_2HPO_4, conductivity values were adjusted to 25, 50, and 100 μS/cm. These combinations of pH and conductivity resulted in a total of 12 different suspending mediums. A multiparameter bench meter HI 255 (Hanna Instruments, Woonsocket, RI, USA) was used to measure pH and conductivity.

Equipment

A high voltage sequencer, model HVS448 (LabSmith, Livermore, CA, USA) was utilized to apply direct current electric fields using platinum-wire electrodes with a diameter of 0.3048 mm (Omega, Stamford, CT, USA). An inverted microfluidic microscope SVM340 (LabSmith, Livermore, CA, USA) with an integrated camera was employed to capture video for μPIV measurements, where a 10X microscope objective was used for all experiments. In order to obtain and analyze the images/videos for PIV measurements, μScope software provided by the microscope manufacturer was employed. A personal computer was required for the operation of the power supply and microscope.

Cell preparation protocols

Syto ® 9 bacterial stain, ex/em 483/503 nm (Molecular Probes, Inc., Eugene, OR) was used to stain *Escherichia coli* (strain Top 10), *Ankistrodesmus sp.* (strain EE98), and *Saccharomyces cerevisiae* (strain ATCC #24858) in order to visualize the cells as required by μPIV. A milliliter of cell culture required three microliters of Syto ® 9, following the incubation for 30 minutes. Excess dye was eliminated by spinning down the cells and washing with deoinized water three times. Each cell sample was resuspended in a buffer solution to an indicated cell concentration: *E. coli* ($6x10^8$ cells/ml), *Ankistrodesmus sp.* ($6x10^7$ cells/ml), and *S. cerevisiae* ($6x10^8$ cells/ml).

Procedure for μPIV experiments

For each experiment, the microchannel (Fig 1) was filled with one of the twelve suspending mediums; then a sample of 80 μL of cell solution was introduced at the inlet reservoir. Electrodes were placed in the inlet and outlet reservoir, pressure-driven flow was eliminated and then the DC electric field was applied. Each experiment was repeated three times in order to obtain reproducibility. Figure 1 shows the region where μPIV measurements were carried out: the central region of the microchannel employing a 256 x 256 pixel interrogation region. With μPIV measurements, EK velocity values (μm/s) were obtained directly.

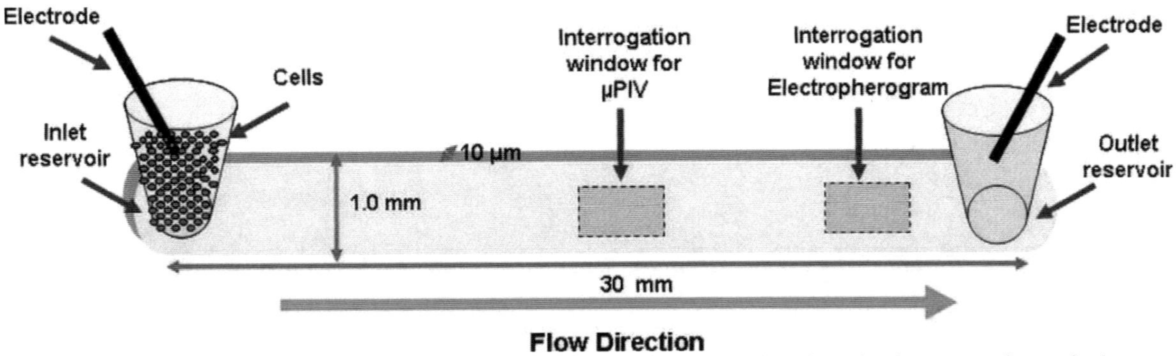

Figure 1. Schematic representation of the experimental set-up, showing the interrogation window employed for μPIV.

Procedure for electropherogram experiments

For electropherogram experiments, a suspending medium with pH of 6 and conductivity of 50 μS/cm was employed to fill the microchannel; afterwards, 200 μl of each cell sample were mixed in the same vial and a sample of 80 μL of the mixed cell solution was introduced at the inlet reservoir of the microchannel. A direct current electric field of 150 V/cm was applied to the microchannel, and during this time a video was recorded by using an interrogation window placed 2.5 cm from the inlet reservoir as shown in Figure 1. The electropherogram generated showed fluorescence measurements vs. time.

Results and Discussion

μPIV characterization

Electrokinetic velocity for bacterial, microalgae, and yeast cells was obtained from μPIV in a glass microchannel as a function of the applied electric field and suspending medium properties (pH and conductivity). For the smaller microorganisms *Escherichia coli* (2 μm) and *Ankistrodesmus sp.* (3 μm), electric fields in the order of 50 and 100 V/cm were used, due to their small size; for the largest cells *Saccharomyces cerevisiae* (5 μm) electric fields up to 300 V/cm were applied since their larger size allowed for good visualization at higher electrokinetic velocities. With EK velocity values (v_{EK}) obtained directly from μPIV, EK mobility values (μ_{EK}) were calculated for each cell with Eqs. (3-5), considering a value of $\eta = 8.91\text{x}10^{-4}$ kg/m·s and $\varepsilon_m = 6.94\text{x}10^{-10}$ F/m. Figures 2a-2c show the complete results of v_{EK} for the three types of microorganisms, where the range of velocity values obtained for applied fields of 50 and 100 V/cm are 198 to 553 μm/s, 194 to 499 μm/s, and 142 to 417 μm/s for *E. coli*, *Ankistrodesmus sp.*, and *S. cerevisiae* respectively. The results of μ_{EK} are shown in Figures 2d-2f for each microorganism, where the range of mobility values are 3.77 to $5.53\text{x}10^{-8}$ m^2/V·s, 3.52 to $5.38\text{x}10^{-8}$ m^2/V·s, and 2.63 to $4.35\text{x}10^{-8}$ m^2/V·s for bacterial, microalgae, and yeast cells, respectively. Confirming the previous results, highest mobility values were obtained from the smaller cells, bacterial, followed by microalgae and yeast cells.

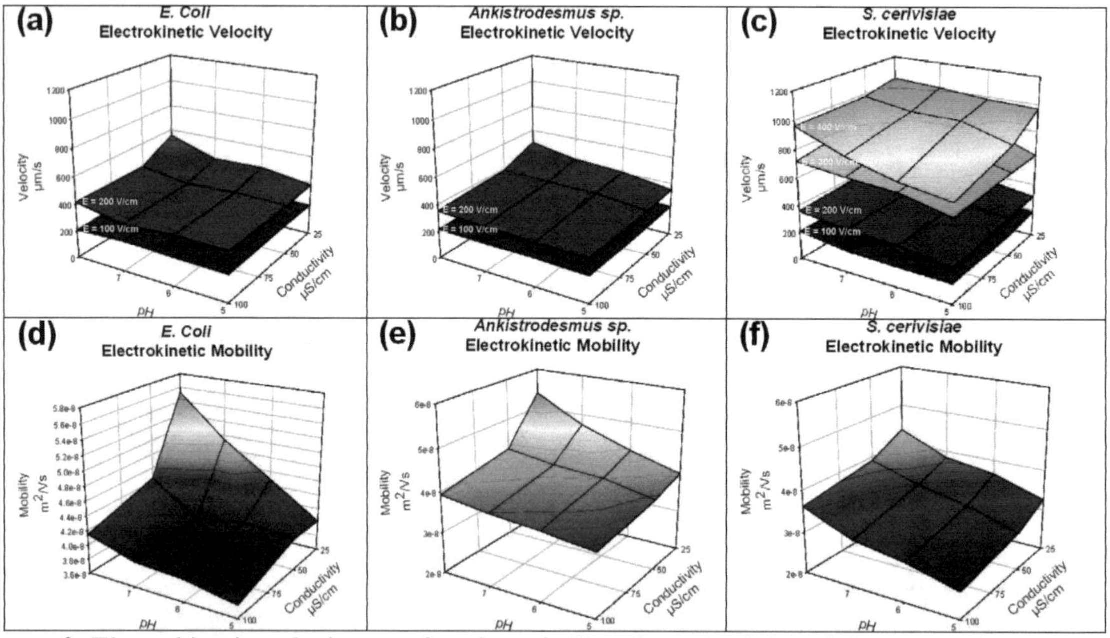

Figure 2. Electrokinetic velocity as a function of pH and conductivity of the suspending buffer of **(a)** *E. coli* **(b)** *Ankistrodesmus sp.* **(c)** *S. cerevisiae.* Electrokinetic mobility as a function of pH and conductivity of the suspending buffer of **(d)** *E. coli* **(e)** *Ankistrodesmus sp.* **(f)** *S. cerevisiae.*

Electropherogram

Experiments were carried out to confirm the presence of different types of microorganisms in a mixture. A sample containing the bacterial, microalgae, and yeast cells was introduced at the inlet reservoir, employing a medium of pH 6 and conductivity of 50 μS/cm. By applying a field of 150 V/cm, cell movement was recorded at an interrogation window for electropherogram shown in Figure 1. Then, the video obtained was analyzed and the fluorescence signal was recorded versus time, by plotting these results an electropherogram was obtained. These results demonstrated that each microorganism arrived at the interrogation window at a different time that was proportional to cell size. The arrival of each microorganism was observed as a "step" in the electropherogram, obtaining as a result "three steps". Table 1 presents the predicted migration times for each microbe at an applied field of 150 V/cm, calculated with the average mobility. Figure 3a shows a schematic representation of the expected results and Figure 3b presents the experimental electropherogram obtained in one minute approximately.

Table 1. Expected migration times to reach the electropherogram interrogation window of the three types of cells at an applied field of 150 V/cm.

Microorganism	Predicted elution time at E=150 V/cm
E. coli	38.1 s
Ankistrodesmus sp.	39.7 s
S. cerevisiae	51.0 s

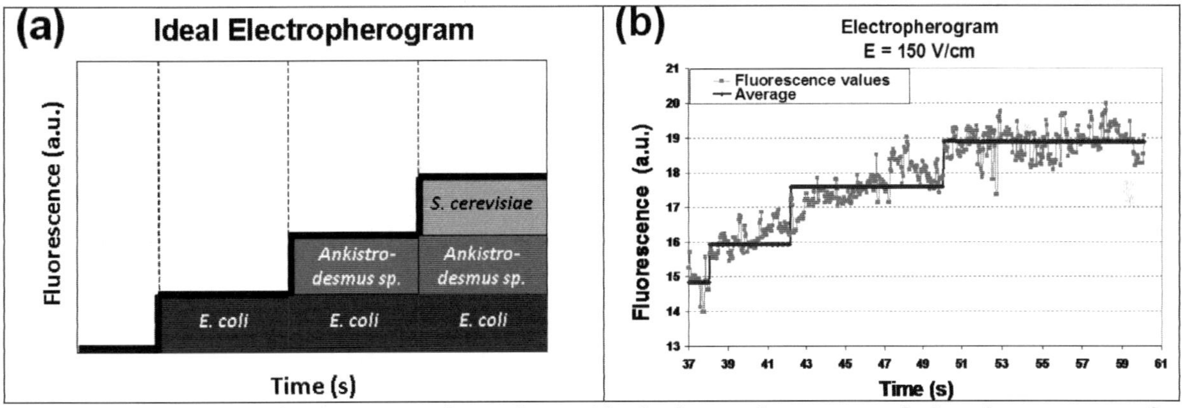

Figure 3. **(a)** Schematic representation of an *ideal* electropherogram of the three types of microorganisms. **(b)** Experimental electropherogram obtained at an applied field of 150 V/cm.

Conclusions

In this study, a simple microscale electrokinetic method was employed in order to differentiate bacterial, microalgae, and yeast cells, applying direct current electric fields in a glass microchannel. Electrokinetic velocity and mobility of the three types of microorganisms were obtained using micro particle image velocimetry. Results showed that suspending medium properties and cell size have a strong effect on electrokinetic migration. Furthermore, the presence of the three types of cells in sample was confirmed with an electropherogram plotted in one minute of processing time, which showed three distinct regions. These results demonstrate that microscale electrokinetic methods have a great potential for fast identification and analysis of microbes such as early detection of pathogens in many food and water safety applications. The results could provide guidelines to aid in the design and selection of operating conditions of micro electrokinetic devices for microbe manipulation and detection.

Acknowledgements

The authors wish to thank the financial support given by CINVESTAV-Monterrey, the Award for Women in Science L'ÓREAL-UNESCO-AMC Mexico 2008, and grant CONACTY-CB-2006-53603.

References

1 C. R. Cabrera, P. Yager, *Electrophoresis* **2001**, *22 (2),* 355.
2 G. M. Whitesides, *Nature* **2006**, *442 (7101),* 368.
3 J. Voldman, *Ann. Rev. Biomed. Eng.* **2006**, *8* 425.
4 D. W. Armstrong, G. Schulte, J. M. Schneiderheinze, D. J. Westenberg, *Anal. Chem.* **1999**, *71 (24),* 5465.
5 H. A. Pohl, Dielectrophoresis, Cambridge University Press, Cambridge, 1978.
6 L. Benguigui, I. J. Lin, *J. Appl. Phys.* **1984**, *56* 3294.
7 B. H. Lapizco-Encinas, M. Rito-Palomares, *Electrophoresis* **2007**, *28 (24),* 4521.
8 J. S. Crane, H. A. Pohl, *J. Electrochem. Soc.* **1968**, *115 (6),* 584.
9 M. P. Hughes, H. Morgan, F. J. Rixon, *Eur. Biophys J. Biophys. Lett.* **2001**, *30 (4),* 268.
10 B. H. Lapizco-Encinas, B. A. Simmons, E. B. Cummings, Y. Fintschenko, *Anal. Chem.* **2004**, *76 (6),* 1571.
11 H. Li, R. Bashir, *Sens. Actuator B-Chem.* **2002**, *86 (2-3),* 215.
12 G. H. Markx, M. S. Talary, R. Pethig, *J. Biotechnol.* **1994**, *32* 29.
13 W. B. Betts, A. P. Brown, *J. Appl. Microbiol.* **1999**, *85* 201S.
14 D. W. Armstrong, L. F. He, *Anal. Chem.* **2001**, *73 (19),* 4551.
15 M. Jackowski, J. Szeliga, E. Klodzinska, B. Buszewski, *Anal. Bioanal. Chem.* **2008**, *391 (6),* 2153.
16 J. I. Martínez-López, H. Moncada-Hernández, J. L. Baylon-Cardiel, S. O. Martínez-Chapa, M. Rito-Palomares, B. H. Lapizco-Encinas, *Anal. Bioanal. Chem.* **2009**, *394 (1),* 293.
17 J. S. Dickson, M. Koohmaraie, *Appl. Environ. Microbiol.* **1989**, *55 (4),* 832.

Dielectrophoretic characterization of microorganisms employing three dimensional carbon electrodes

Héctor Moncada-Hernández,[1] *Victor H. Pérez-González,*[1] *Rodrigo Martinez-Duarte,*[2] *Sergio O. Martínez-Chapa,*[1] *Marc J. Madou,*[2] *Blanca H. Lapizco-Encinas*[*,3] *Email:blapizco@cinvestav.mx*

[1] *BioMEMS Reseach Chair, Tecnológico de Monterrey, Monterrey NL, Mexico.*
[2] *Mechanical & Aerospace Engineering, University of California, Irvine, Irvine, CA*
[3] *Microscale Bioseparations Laboratory, CINVESTAV-Monterrey, Apodaca NL, Mexico.*

Abstract

Dielectrophoresis is electrokinetic transport mechanism that has been successfully employed to manipulate particles in microfluidic devices. Several applications have been demonstrated using planar metallic electrodes or insulating obstacles; which until now had been the two main trends on dielectrophoretic devices. The use of 3D electrodes in DEP applications has attractive advantages: since it allows for higher separation throughputs and requires lower electric potentials than the traditional planar electrodes or insulating structures approaches. Carbon-based-electrode Dielectrophoresis (CarbonDEP) is a novel technique that employs 3D carbon structures or surfaces as electrodes with advantages that make it a more suitable material than those traditionally used in metal-electrode DEP. One of the most promising techniques for carbon microfabrication is that of **C**arbon **M**icro**E**lectro**M**echanical **S**ystems (C-MEMS). Presented here is the implementation of a bioparticle characterization platform based on CarbonDEP. A microdevice containing arrays of 3D carbon electrodes was employed to manipulate carboxylated polystyrene particles and yeast cells in order to characterize cell dielectric properties. Cells responses were obtained by varying the magnitude and frequency of the applied AC potential as well as the characteristics of the suspending medium. Simulation work allowed building a model to predict cells responses. These results have the potential to be used as guidelines for the design and operation of CarbonDEP-based systems. Potential applications include clinical analyses, environmental screening for water contamination, and optimization of cell culture techniques, improvement of clean energy production methods and food safety methodologies and procedures.

Introduction

Most dielectrophoretic devices are developed based on the metal electrode scheme, in which planar electrodes are etched on the microchannel floor. These planar electrodes will then produce the non-uniform electric field required to induce an electric dipole moment on the particles of interest. This scheme has some important drawbacks. The dielectrophoretic force will act only in the particles located in the vicinity of the planar electrodes, making it necessary to loop the process several times damaging the particles. Other important drawbacks of conventional DEP methods are sample electrolysis and high fabrication costs. The other leading trend in DEP research is insulator-based DEP (iDEP), where an electric potential is applied across the complete channel giving rise to a uniformly distributed electric field. Along the channel, insulating posts or hurdles deform the field giving rise to large electric field gradients where DEP occurs.[1] This technique also presents an inconvenient characteristic. Joule heating occurs due to the large electric potential needed to create high electric field gradient zones.[2]

Applications of DEP span a wide range of interest areas in current research. DEP has been successfully implemented to either separate,[3, 4] concentrate, transport,[5] detect,[6] or determine dielectric properties of particles.[7-9] The work presented herein treats the problem of dielectric properties determination of particles using CarbonDEP, a novel technique that brings together the advantages of iDEP and electrode-DEP. CarbonDEP relies on 3D carbon electrodes to create electric field gradients necessary to observe a dielectrophoretic force. Carbon DEP exhibits several advantages over the two main DEP trends, it affects the complete bulk, not only particles near the floor of the channel; reduces the electric potential used as excitation signal minimizing electrolysis and joule heating. Also, carbon DEP presents excellent biocompatibility, is chemically inert and has good mechanical properties.[8]

This work presents a novel technique to determine the dielectric properties of particles. A microchannel with an array of conductive posts was excited with alternating current (AC) electric potentials. The frequency of the excitation signal was varied until the DEP force was zero. With this frequency value a complex mathematical system was solved to obtain the equivalent conductivity and permittivity of the particles of interest. Finally using these values, separation was performed.

Theory of Dielectrophoresis.

Dielectrophoresis is a technique developed by late Prof. Herbert Pohl in 1951.[10] It describes the polarization effect of a particle suspended in an electrolytic solution when subjected to a nun uniform DC or AC electric field.[11-14] The dielectrophoretic force experienced by a spherical particle is defined as:

$$\mathbf{F}(\omega) = 2\pi\varepsilon_m r^3 \operatorname{Re}\left[f_{CM}\right]\nabla E_{rms}^2 \tag{1}$$

where ε_m is the permittivity of the suspending medium, r is the particle radius, and f_{CM} is the Clausius-Mossotti (CM) factor, which is defined as:

$$f_{CM} = \frac{\varepsilon_p^* - \varepsilon_m^*}{\varepsilon_p^* + 2\varepsilon_m^*} = \frac{\sigma_p^* - \sigma_m^*}{\sigma_p^* + 2\sigma_m^*} \tag{2}$$

where $\varepsilon^* = \varepsilon - j\sigma/\omega$ and $\sigma^* = \sigma + j\varepsilon\omega$ refer to the complex permittivity and complex conductivity respectively. In (2), subscripts p and m refer to particle and medium respectively, and ω is the angular frequency. The CM factor is a measure of the ease of polarizability of a particle in a given suspending medium. When the value of the CM factor is positive, it indicates that the DEP force experienced by the particle will be positive or attractive. Otherwise, if f_{CM} is negative, the force will be negative or repulsive.

When the particles of interest are covered with shells of different dielectric properties (Figure 1), the equivalent values of these properties must be calculated. For this, a simple idealized multishell model was used. The equivalent complex permittivity of a multishell particle is defined as:[13]

$$\varepsilon_n^{'*} = \varepsilon_n^* \left\{ \frac{a^3 + 2\left(\dfrac{\varepsilon_{n+1}^* - \varepsilon_n^*}{\varepsilon_{n+1}^* + 2\varepsilon_n^*}\right)}{a^3 - \left(\dfrac{\varepsilon_{n+1}^* - \varepsilon_n^*}{\varepsilon_{n+1}^* + 2\varepsilon_n^*}\right)} \right\} \tag{3}$$

where $a = r_{n-1}/r_n$. Finally, the value obtained with (3) will be used in (2) to compute the value of f_{CM}.

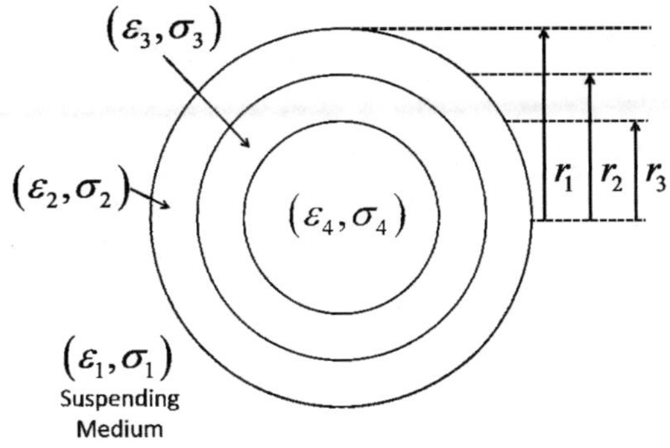

Figure 1. Schematic diagram of a particle with three layers.

Finally, the crossover frequency of the CM factor is uniquely defined by (4) as a function of the particle and buffer solution dielectric properties.[15] This value represents the frequency of the excitation signal at which the particle under analysis will experience a null dielectrophoretic force.

$$f_X = \frac{1}{2\pi\varepsilon_0}\left(\frac{\sigma_p^2 + \sigma_p\sigma_m - 2\sigma_m^2}{\varepsilon_p^2 + \varepsilon_p\varepsilon_m - 2\varepsilon_m^2}\right)^{1/2} \tag{4}$$

Equation (4) can be mathematically derived by equating the Real part of the CM factor to zero, and solving for the frequency. However, there is a modification that must be performed on the expression in order to extract the crossover frequencies of interest. For a given particle in a specific suspending medium, a crossover frequency exists only when either one of two condition is satisfied. The first condition is satisfied when $\sigma_p > \sigma_m$ and $\varepsilon_p < \varepsilon_m$ giving rise to a crossover frequency from positive DEP to negative DEP. The second condition is the exact opposite, that is when $\sigma_p < \sigma_m$ and $\varepsilon_p > \varepsilon_m$ giving rise to a crossover frequency from negative DEP to positive DEP. When any of these conditions occur, the result given by evaluation of Equation (4) is a complex frequency with a dominant imaginary part. Therefore, in this work, the crossover frequency equation has been redefined as:

$$f_X = \frac{1}{2\pi\varepsilon_0}\left(\frac{\sigma_p^2 + \sigma_p\sigma_m - 2\sigma_m^2}{2\varepsilon_m^2 - \varepsilon_p\varepsilon_m - \varepsilon_p^2}\right)^{1/2} \tag{5}$$

Mathematical simulation

The performance of a Carbon-DEP microfluidic device is modeled in the present work. Modeling is a fundamental part of the present work since it allows optimizing the experimental phase. If no mathematical or simulation model exists, obtaining adequate results in the laboratory would be tedious and difficult. The results obtained via computational simulation provide guidelines for setting up experiments.

In this work, dielectrophoretic phenomena have been studied through a Finite Element Analysis (FEA) method. FEA has been implemented with the computational software COMSOL Multiphysics©. In addition, MATLAB (The Mathworks) was used to encounter the crossover frequency of homogeneous particles as a function of particle and buffer solution dielectric properties, as well as for the estimation of dielectric properties of the particles.

The geometry of a Carbon-DEP microfluidic device was captured with COMSOL Multiphysics©. The microchannel consisted of two arrays of 20 x 5 electrodes (carbon posts) each (Fig. 2a). The posts lie above interdigitated planar electrodes in a chess board arrange, used both to provide an excitation signal to the posts, and to be used as DEP traps. These interdigitated electrodes are connected to an external excitation source (Fig. 2c). Posts are 70 μm high with a 12.5μm radius and a center to center separation of 45 μm and 100 μm in the X and Y axis respectively.

The approach taken in this work is that of finding the crossover frequency f_{X1} of an unknown particle when suspended a buffer solution with dielectric properties ε_{m1} and σ_{m1}, as well as f_{X2} of the same unknown particle when suspended a buffer solution with dielectric properties ε_{m2} and σ_{m2}. Substituting these values into equation (4), two equations are obtained that have ε_p and σ_p as unknown variables. By solving these two resultant equations simultaneously, a set of four permittivity-conductivity solution pairs are calculated, from which only one these pairs show a positive value for both properties.

Experimental Set-up.

A Carbon DEP device with the characteristics described above was fabricated by pyrolysis of SU-8 structures defined by a two-step photolithography process following standard C-MEMS techniques. The electrodes were coated with a thin SU-8 layer in order to protect the connecting leads and provide a planar surface for the microchannel; which was made from a double sided pressure sensitive adhesive. The microchannel was then aligned with a previously drilled (inlet and outlet reservoir) polymer cover.

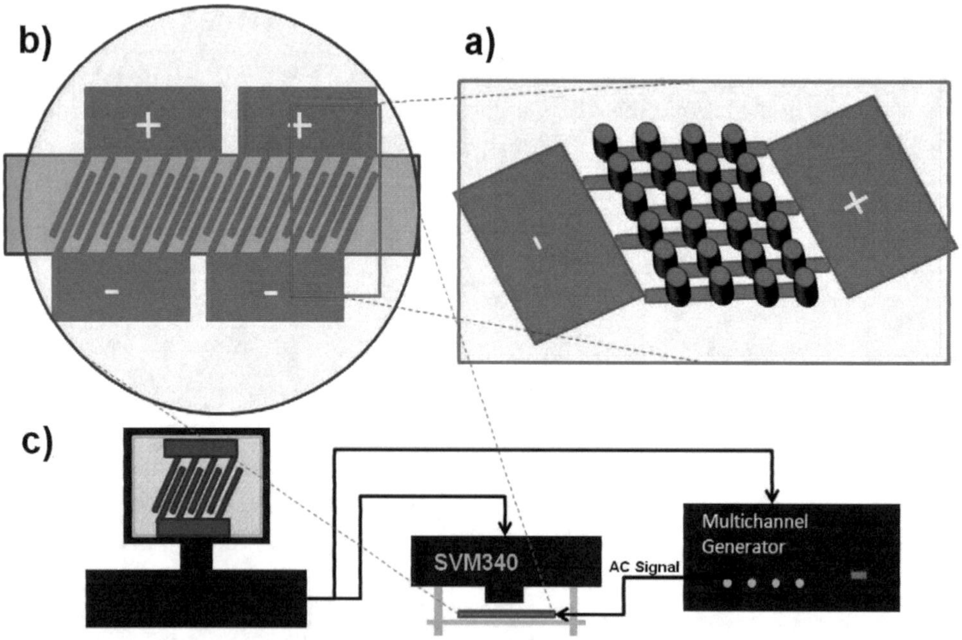

Figure 2. Experimental Set-up. a) representation of microchannel and carbon cylindrical electrodes. b) top view of microdevice. c) Communication and control between the inverted microscope and the multichannel generator with a personal computer.

Buffer solutions were prepared from DI water, adjusting the conductivity to 50 and 200 μS/cm with K_2HPO_4. Conductivities were measured with a multiparameter bench meter HI 255 (Hanna Instruments, Woonsocket, RI, USA).

Saccharomyces cerevisiae cells (ATCC 24858) were cultured in YM broth for 18 hr at 30° C in a shaker to a cell concentration of $6x10^6$ cell/mL (late log phase) verified by OD measurements at 600 nm. Cells were then washed with DI water to remove the culture media and resuspended in DI water to be labeled with the fluorescent dye Syto® 9 (Invitrogen, Carlsbad, CA) that fluoresces green (excitation/emission 483/503 nm) and finally resuspended in the respective buffer solutions to a concentration of $6x10^7$ cells/mL. Carboxylated polystyrene particles with a diameter of 10.14 μm (Invitrogen, Carlsbad, CA) were diluted to a concentration of $4.2x10^6$ spheres/mL in the buffer solutions previously mentioned.

The sample was introduced to the microdevice employing a micropipette. The microdevice was mounted under an inverted epifluorescence video microscope for microfluidics SVM340 (LabSmith, Livermore, CA, USA) turned upside down. Electric signal was provided to the microdevice by a high frequency multichannel generator (Tecnológico de Monterrey, Monterrey, NL, México) ranging from 1 to 20 Vpp and up to 20 MHz as shown in Figs. 2b and 2c. A personal computer is employed to manipulate the communication and operation of the microscope and the multichannel generator (Fig. 2c).

Results

It is important to justify the use of C-MEMS technology for the fabrication of particle characterization devices. Figure 3 shows a plot of the gradient of the squared electric field over one section of the channel. It can be observed in this plot that the magnitude of this gradient value is greatly increased in the vicinity of the carbon posts, thus increasing the throughput of particle trapping, which is essential to perform the dielectric characterization of unknown particles.

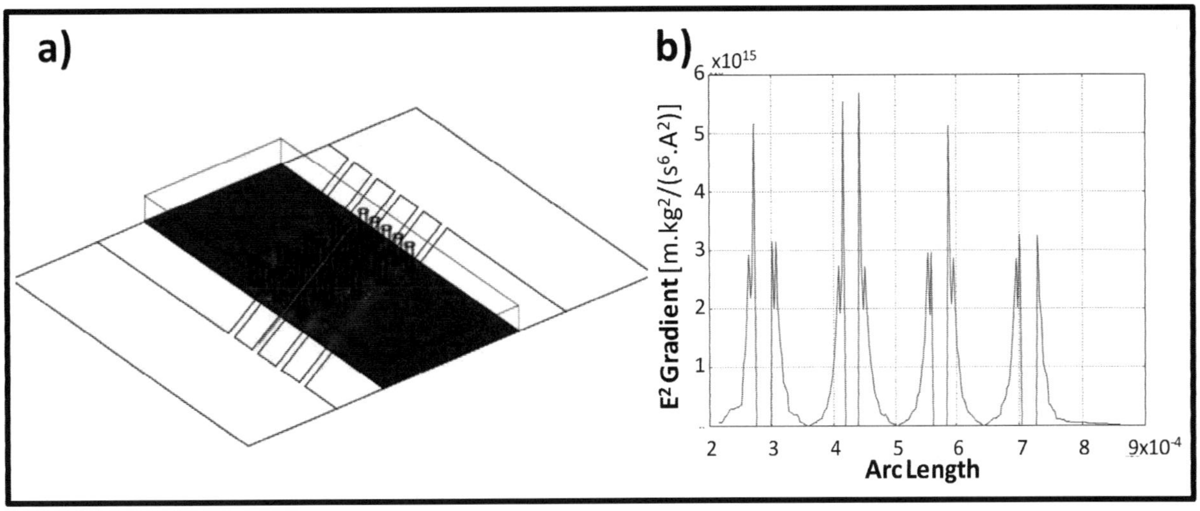

Figure 3. a) Slice plot of the distribution of ∇E_{rms}^2 throughout the device at a height of 10μm. b) Graph of the magnitude of ∇E_{rms}^2 along the red line plotted on a).

Throughout mathematical simulations we have validated the efficiency of the model proposed in this work. By using a set of arbitrary values of particle dielectric properties, and modeling the buffer solution on which the particles were to be suspended; it was possible to determine a set of crossover frequencies of these randomly selected virtual particles using Equation (5). Figure 4 shows crossover frequencies for different sets of homogeneous particle properties.

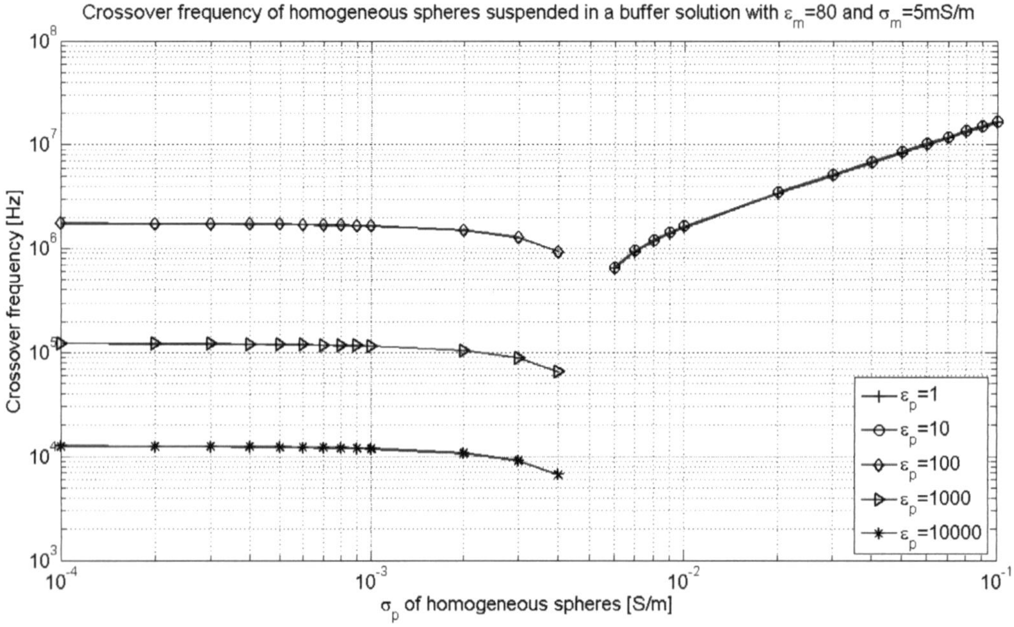

Figure 4. Crossover frequency of arbitrary homogeneous particles.

Figure 4 exhibits the fact that the particle permittivity turns out to be a determinant parameter while calculating the crossover frequency. From the plot, it can be observed that with values of particle permittivity around and below that of the medium permittivity, the crossover frequency increases with the permittivity, but after a certain value of permittivity is reached, the crossover frequency decreases with increments of particle permittivity. It can also be observed that two behaviors are shown in Figure 4; from left to right there are three lines, these lines represent crossover frequencies from negative DEP to positive DEP. From right to left there are two lines that practically exhibit the same behavior. These lines represent crossover frequencies from positive DEP to negative DEP.

The crossover frequencies determined were then used as input to the dielectric properties estimator described in the mathematical modeling section above. The results obtained throughout this procedure demonstrated that, in the absence of measurement error, the model is able to precisely extract the dielectric properties of homogenous particles. The next attempt was to estimate dielectric properties of multilayered cells. In order to perform this parameter extraction, the shell model was used to take into account the complete dielectrophoretic effect experienced by each layer of the cell.

Experimental results were determined by observing changes between positive and negative DEP of particles. In Figure 5 cases of positive and negative DEP of the polystyrene particles in the microdevice under frequencies ranging from 0.5 kHz to 10 kHz can be observed. Frequencies for positive (particles attached to the electrodes) and negative (particles repelled from the electrodes) dielectrophoresis are 0.6

kHz and 20 kHz, respectively. Analogous with the spheres, yeast cells where submitted to frequencies ranging from 1 kHz to 100 kHz.

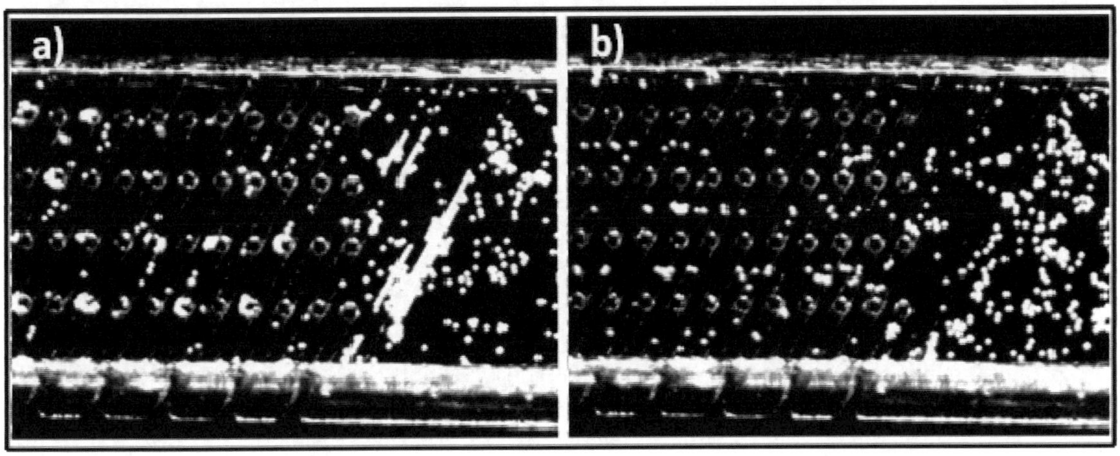

Figure 5. Dielectrophoretic manipulation of carboxylated polystyrene particles in a suspending medium of 200 uS/cm. and 15 Vpp. a) Particles showing positive DEP at a signal of 0.6 kHz. b) Particles showing negative DEP at a signal of 20 kHz.

Estimation results in this line are not completely satisfactory yet. Figure 6 shows the error of estimation of the model when a noise-free frequency measurement is considered for viable yeast cells.

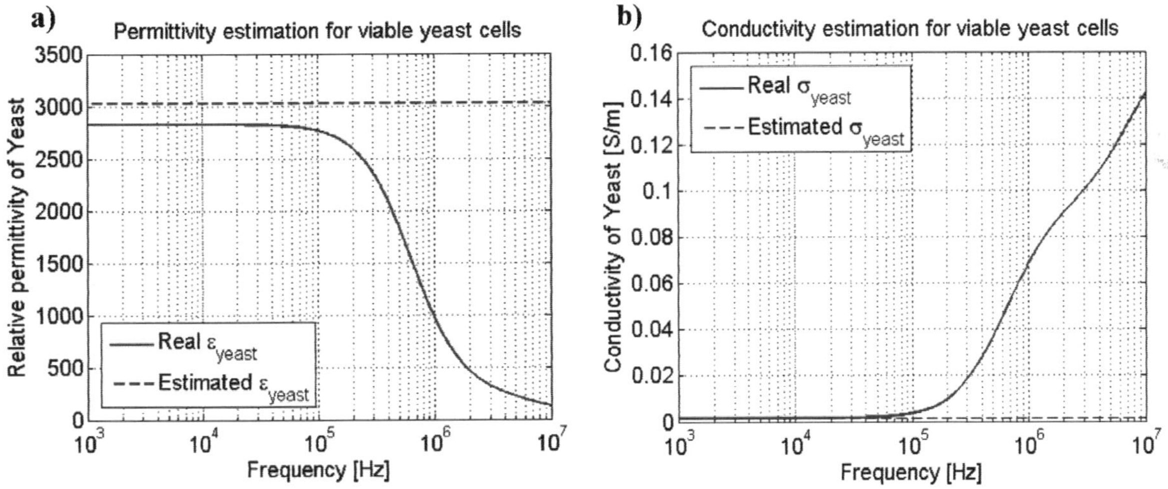

Figure 6. Comparison between the estimated and real values of: a) the equivalent permittivity of yeast cells and b) the equivalent conductivity of yeast cells.

From the results plotted on Figure 6 it can be seen that the estimator is capable of obtaining values that are close to those of the yeast cells. But this is only true for a limited frequency range.

Conclusions

The potential of carbon-electrode-based dielectrophoresis with mathematical modeling tools is proven to be latent in this work. Positive and negative dielectrophoresis of polystyrene and viable yeast cells was demonstrated in order to determine their crossover frequencies. These crossover frequencies data along with the conditions of the suspending media served as an input to the mathematical model, which provides an estimation of the particle dielectric properties. There is still work to do to improve the mathematical model performance for a multishell particle. It is also important to perform a statistical characterization of the estimated values, this in order to understand the negative effect of inaccurate frequency measurements. Visual determination of crossover frequencies is also to be optimized to determine more accurate values of frequencies for each particle type.

References

1. E. B. Cummings and A. K. Singh, *Analytical Chemistry*, 2003, **75**, 4724-4731.
2. P. Sabounchi, D. E. Huber, M. P. Kanouff, A. E. Harris and B. A. Simmons, Joule heating effects on insulator-based dielectrophoresis, in The 12th International Conference on Miniaturized Systems for Chemistry and Life Sciences (MicroTAS 2008), San Diego CA, 2008, vol. 1, pp. 50-52.
3. H. Shafiee, M. B. Sano, E. A. Henslee, J. L. Caldwell and R. V. Davalos, *Lab on a Chip*, 2010, **10**, 438-445.
4. M. Urdaneta and E. Smela, *Electrophoresis*, 2007, **28**, 3145-3155.
5. U. Lei, C. W. Huang, J. Chen, C. Y. Yang, Y. J. Lo, A. Wo, C. F. Chen and T. W. Fung, *Lab on a Chip*, 2009, **9**, 1349-1356.
6. G. Medoro, N. Manaresi, A. Leonardi, L. Altomare, M. Tartagni and R. Guerrieri, A lab-on-a-chip for cell detection and manipulation, 2002.
7. K. L. Chan, H. Morgan, E. Morgan, I. T. Cameron and M. R. Thomas, *Biochimica et Biophysica Acta-Molecular Basis of Disease*, 2000, **1500**, 313-322.
8. R. Martinez-Duarte, R. A. Gorkin, K. Abi-Samra and M. J. Madou, *Lab on a Chip*, 2010, **10**, 1030-1043.
9. C. F. Chou and F. Zenhausern, *Ieee Engineering in Medicine and Biology Magazine*, 2003, **22**, 62-67.
10. H. A. Pohl, *Dielectrophoresis : the behavior of neutral matter in nonuniform electric fields*, Cambridge University Press, Cambridge ; New York, 1978.
11. N. G. Green, A. Ramos and H. Morgan, *Journal of Physics D-Applied Physics*, 2000, **33**, 632-641.
12. H. Morgan and N. G. Green, *AC electrokinetics : colloids and nanoparticles*, Research Studies Press, Baldock, Hertfordshire, England, 2003.
13. T. B. Jones, *Electromechanics of particles*, Cambridge University Press, Cambridge ; New York, 1995.
14. T. B. Jones, *IEEE Engineering in Medicine and Biology Magazine*, 2003, **22**, 33-42.
15. I. Ermolina and H. Morgan, *Journal of Colloid and Interface Science*, 2005, **285**, 419-428.

Assessing particle Selectivity of an insulator-based dielectrophoretic microdevice

Ana V. Chávez-Santoscoy,[1] Javier L. Baylon-Cardiel,[1] Héctor Moncada-Hernández[1] and Blanca H. Lapizco-Encinas[2,] Email: blapizco@cinvestav.mx*

[1] BioMEMS Reseach Chair, Tecnológico de Monterrey, Monterrey NL, Mexico.
[2] Microscale Bioseparations Laboratory, CINVESTAV-Monterrey, Apodaca NL, Mexico.

Abstract

Miniaturization has brought important advantages to separation technologies; such as reduced sample and reagent consumption; higher resolution and sensitivity, low cost and shorter processing time. Dielectrophoresis (DEP) is an extensively employed micro analytical technique with great potential for the manipulation of a wide array of bioparticles. Insulator-based DEP (iDEP) is a dielectrophoretic mode where nonuniform electric fields are created employing insulating structures.
In this work, an evaluation of the selectivity of an iDEP microdevice by employing mixtures of polystyrene nano and microparticles is presented. Experimental and mathematical modeling work was carried in order to assess the selectivity of an iDEP glass microdevice to immobilize only the large particles from a binary mixture. The effects of particle concentration, concentration ratio, size ratio, and magnitude of the applied electric potential (200 to 600 V) on the microdevice selectivity were studied. The results demonstrated that high selectivity can be obtained with iDEP.

Introduction

Microfluidics has significantly advanced the development of separations technologies. It offers significant advantages over traditional bench-scale equipments, such as: small quantities of samples and reagents required, higher resolution and sensitivity obtained, and microdevices are lower cost and require shorter processing times [1]. Dielectrophoresis (DEP) has become one of the most important micro analytical techniques, since it can be applied for the manipulation, separation and concentration of macromolecules, virus, cell organelles, bacterial, yeast and parasites. DEP refers to the interaction of a dielectric particle with a spatially non-uniform electric field where the particle becomes polarized. Movement induced in a particle depends on the characteristics of the electric field and the dielectric properties of the particle and suspending medium. When the electric field is spatially non-uniform, the dipolar charges and the electric field, produce a net movement of the particle in the direction to the higher field magnitude. Since particle motion is independent on electric field polarity, DEP can take place in alternating current (AC) or direct current (DC) electric fields [6].
Traditionally, DEP studies have been carried out employing arrays of microelectrodes in order to generate spatially non-uniform electric fields. However, the use of microelectrodes involves high cost of electrode construction, complex fabrication processes and a decrease of functionality due to fouling effects [2,8]. An alternative to this technique is the insulator-based DEP (iDEP) approach; where an electric field is applied across an array of insulating structures in order to create regions of higher and lower field strength [8-12].
The present work is focused on the evaluation of the selectivity of a glass iDEP microdevice by employing binary mixtures of polystyrene nano and microparticles. Experiments employing iDEP as well as mathematical modeling with *COMSOL Multiphysics* were carried out in order to observe and predict the response of the particles to assess the selectivity of the microdevice. Results demonstrated selective particle immobilization from a binary mixture, allowing for successful particle sorting and separation. The findings reported here demonstrate the great potential of iDEP as technique for continuous, highly-selective mixture fractionation.

Experimental

Setup

Experiments were conducted in a microfluidic device consisting of two circular glass wafers. The top glass wafer contained drilled holes that transverse the entire depth of the wafer and were employed as liquid reservoirs. The bottom glass wafer contained patterned microchannels fabricated with standard wet-etching techniques. The microchannel employed in this study is 20-μm deep, 10.16-mm long and 1-mm wide, containing an array of 40 cylindrical posts arranged in 4 columns of 10 posts each. The posts have a diameter of 200 μm, a height of 20 μm and were in a square array spaced 250 μm center-to-center, as shown in Figure 1.

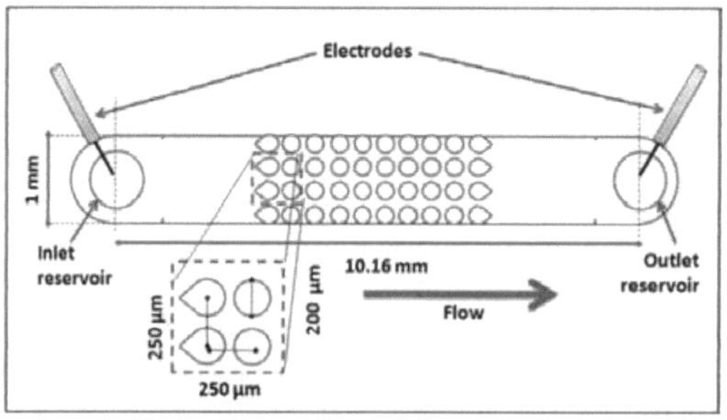

Figure 1. Schematic representation of experimental setup showing top view of microchannel.

A high voltage sequencer was used to apply a DC electric potential (model HVS448, LabSmith, Livermore, CA). The response of the particles was recorded in the form of videos by employing an inverted epifluorescence video microscope for microfluidics, model SVM340 (LabSmith, Livermore, CA).

Particles

The particles employed for this work were polystyrene carboxylated-modified fluorescent green-yellow microspheres (ex/em 505/515 nm), radius of 100-nm and 500-nm (Invitrogen, Carlsbad, CA, USA). Also, polystyrene carboxylated-modified fluorescent suncoast-yellow microspheres (ex/em 580/600 nm) with radius of 1.14 μm and flash red microspheres (ex/em 660/690 nm) with radius of 195 nm were employed (Bangs Laboratories, Fishers, IN, USA).

Experimental procedure for dielectrophoresis experiments

Binary mixtures of microspheres using different particle concentration values and concentration ratios were employed. Table 1 describes size combinations of microspheres employed, including microspheres concentration and concentration ratio. Microspheres solutions were prepared employing distilled water with pH = 6.0 and conductivity of 2.45 μS cm^{-1}.

Mathematical model

For simulation work, a 2-dimensional model was built. The depth of the device was not taken into account, since the channel depth is too small compared to its length, and the change of the electric field in this direction is negligible. The Laplace equation:

$$\nabla^2 \phi = 0 \qquad (1)$$

is solved for the geometry considered, using the finite element method, as implemented with *COMSOL Multyphysics* software.

Table 1. Description of systems (binary mixtures) of microspheres employed for experiments.

System label	Microspheres size (radius)	Particle concentration $C_{p,s}$ - $C_{p,l}$ (microspheres mL^{-1})	Concentration ratio ($C_{p,s}$: $C_{p,l}$)
1-1a	195 nm – 500 nm	1 x 10^9 – 1 x 10^8	10:1
1-1b		1 x 10^{10} – 1 x 10^9	
1-2a	195 nm – 500 nm	1 x 10^9 – 2 x 10^6	500:1
1-2b		1 x 10^{10} – 2 x 10^7	
1-3a	195 nm – 500 nm	1 x 10^9 – 1 x 10^6	1000:1
1-3b		1 x 10^{10} – 1 x 10^7	
2-1a	100 nm – 1.14 µm	1 x 10^9 – 1 x 10^8	10:1
2-1b		1 x 10^{10} – 1 x 10^9	
2-2a	100 nm – 1.14 µm	1 x 10^9 – 2 x 10^6	500:1
2-2b		1 x 10^{10} – 2 x 10^7	
2-3a	100 nm – 1.14 µm	1 x 10^9 – 1 x 10^6	1000:1
2-3b		1 x 10^{10} – 1 x 10^7	
3-1a	500 nm - 1.14 µm	1 x 10^7 – 1 x 10^7	1:1
3-1b		1 x 10^8 – 1 x 10^8	
3-2a	500 nm - 1.14 µm	1 x 10^7 – 2 x 10^6	5:1
3-2b		1 x 10^8 – 2 x 10^7	
3-3a	500 nm - 1.14 µm	1 x 10^7 – 1 x 10^6	10:1
3-3b		1 x 10^8 – 1 x 10^7	

Motion of the particles

Particle transport along the micro-channel is assumed to depend on electrokinetic (EK), dielectrophoretic mechanisms and particle size [9,22,23]. The particle flux j can be expressed in terms of an electrokinetic velocity (u_{ek}) and dielectrophoretic (u_{dep}) velocities as:

$$\vec{j} = C(\vec{u}_{ek} + \vec{u}_{dep}) \quad (2)$$

$$\vec{u}_{ek} = \mu_{ek}\vec{E} \quad (3)$$

$$\vec{u}_{dep} = -\mu_{dep}\nabla E^2 \quad (4)$$

where C is the concentration of particles, u_{ek} and u_{dep} are the electrokintic and dielectrophoretic velocities, respectively. For a spherical particle of radius r_p, the dielectrophoretic mobility μ_{dep} can be is expressed as [22]:

$$\mu_{dep} = \frac{\pi r_p^{\ 2} f_{CM}\varepsilon_m}{3\eta} \quad (5)$$

where ε_m is the permittivity of the suspending medium, η is the viscosity of the suspending medium and f_{CM} is the Clausius-Mossotti factor. In this work, only DC electric fields were employed therefore the f_{CM} can be expressed employing the real conductivities:

$$f_{CM} = \frac{\sigma_p - \sigma_m}{\sigma_p + 2\sigma_m} \quad (6)$$

In this study, the conductivity of polystyrene particles was considered negligible (σ_p=0 µS cm^{-1}) [25]

compared to that of the suspending medium. This assumption was supported by calculations of f_{CM} employing σ_p values reported in the literature [25] where all values for f_{CM} were very close to -0.5.

Dielectrophoretic trapping

Trapping of particles occurs when the contribution of DEP to the particle flux overcomes the contribution of EK. Therefore, regions of dielectrophoretic trapping along the microchannel can be developed where the total particle flux j along the electric field lines is zero. By combining eqns. (2), (3) and (4), the condition for trapping due to DEP for a spherical particle is:

$$\frac{\mu_{dep}\nabla E^2}{\mu_{ek}E^2}\cdot\vec{E}>1 \qquad (7)$$

When relation (7) is satisfied in a region along the channel, particles are immobilized due to DEP.

Results and discussion

A set of experiments was carried out to assess the selectivity of the microdevice for the separation of binary mixtures. Mathematical modeling was performed in order to evaluate the potential of simulation work for the prediction of the dielectrophoretic response of the particles and the system selectivity. The effect of several parameters on the selectivity of the iDEP microdevice was analyzed [23].

Model results

By solving Eqn. (1), it is possible to estimate the distribution of the electric potential ϕ along the array of cylindrical insulating structures. The model, built using *COMSOL Multiphysics* software, allows the estimation of the electric field E, the gradient of E^2, and the locations of regions where dielectrophoretic trapping occurs (Eqn. 7). Figure 2a shows the distribution of the electric field when an electric potential ϕ = 350 V is applied across the microchannel. The areas of higher field intensity are located at the constrictions between the posts. DEP trapping regions are greatly related to the gradient distribution on the x-direction, which is shown in Figure 2b. There are two opposite regions with the same shape where the magnitude of the x-component takes its limiting values. From left to right, the first region (red color) is for the positive part as the gradient approaches the constriction, and the second region (blue color) is for the negative part as the gradient moves away after the constriction. It would be expected that a flow of particles dominated by DEP diverges at the enclosed region between the negative and positive part of the gradient, causing the immobilization of particles.

Magnitude of electric potential required for selective trapping

Figure 3a shows that for a given system and its subsystems, the magnitude of the electric potential required to immobilize the large particles does not vary significantly as a function of the concentration ratio or particle concentration. Comparing different systems, one can observe that there is a variation to be considered among different systems. For example, we compare systems 2 and 3, where the large particles in all these systems are the 1.14-µm-radius microspheres. In subsystems 2 (2-1, 2-2 and 2-3) the average required potential is 333 V; while the average potential for system 3 is 260 V. It can be seen that by increasing the size of the smaller particle in binary mixture, the magnitude of the required potential to achieve selective immobilization of the larger particles decreases.

Particle path lines were simulated employing the mathematical model, when a pathline ends at the edge of an insulating post, the particle is expected to be immobilized due to DEP overcoming any other transport mechanisms. Figures 3b to 3d show particles pathlines obtained by the superposition of dielectrophoretic and electrokinetic velocities; these pathlines describe the particle motion through the array of insulating structures. The pathlines are calculated from the numerical solution of Eqn. (1), using Eqn. (3) and Eqn. (4) to compute the velocities. Red lines represent the trajectories of the large particles of a mixture, and blue lines represent the small particles. Figure 3b show the pathlines for systems 1-1 to 1-3, applying the average potential (ϕ= 550V) observed for selective trapping from experimental results as reported in Fig. 3a. Large particles are expected to be immobilized between the insulating posts, since their trajectories end at the posts, while the small particles are expected to keep flowing between posts. Figure 3c shows particle pathlines for systems 2-1 to 2-3 with ϕ = 350 V where small particles are moving and large particles have their pathlines ended at the edge of the insulating posts. In Figure 3d, the pathlines for the systems 3-1 to 3-3 are presented with ϕ = 300 V, showing similar behavior Results in Figures 3b-3d demonstrate that particle immobilization can be predicted studying particle pathlines.

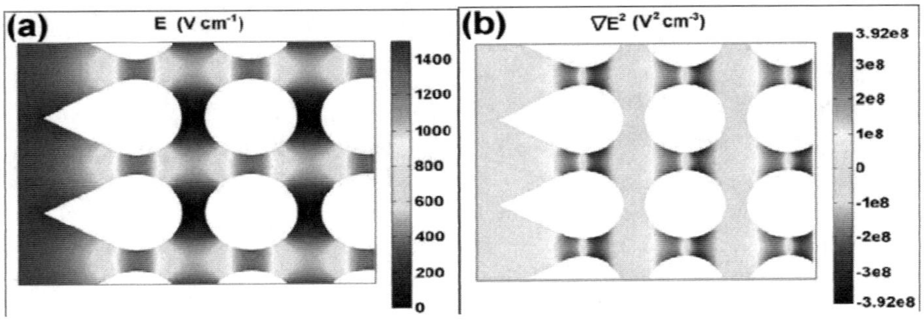

Figure 2. (a) Electric field distribution when a potential of 350 V is applied. (b) Distribution of the x component of the gradient of E^2 over the insulating structures array, showing a region for positive (light) and negative (dark) gradient.

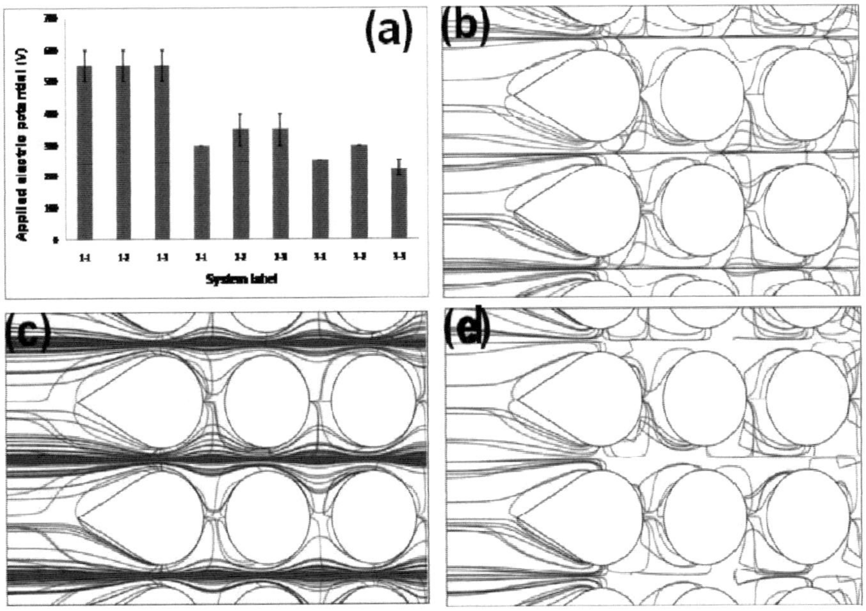

Figure 3. (a) Effect of particle concentration and particle concentration ratio over the magnitude of the electric field required for selective dielectrophoretic immobilization of the large particles of each binary system. Velocity pathlines for (b) systems 1-1 to 1-3, (c) systems 2-1 to 2-3 and (d) systems 3-1 to 3-3. Blue pathlines represent the small particles of the mixture, while red pathlines represent the large particles.

Shape and size of regions of dielectrophoretic immobilization of particles

The effect of particle concentration and particle concentration ratio was also assessed in this study. Figure 4 presents the selective trapping of the large particles in systems 2-1, 2-2 and 2-3. When the ratio of small-to-large particle concentrations is in the order of 10:1 ($C_{p,s}$:$C_{p,l}$), as in mixtures 2-1 applying a potential ϕ = 300 V, it is observed that the large particles tend to concentrate in the two regions where the gradient of E^2 is greater, close to the insulating posts. In contrast, small particles in these systems are observed to move along the channel under the influence of the electrokinetic velocity (Fig. 4a). This shape of the trapping regions suggests that the flow of small particles is very strong. On the other hand, for mixture 2-2b employing a particle concentration ratio of 500:1, large particles trapped due to DEP are immobilized in the shape of short "*pearl-chains*" forming "*lines*", as shown in Figure 4b for $\phi = 400$ V. In this case, the flow of small particles seems to have an important effect over the shape of the trapping region.

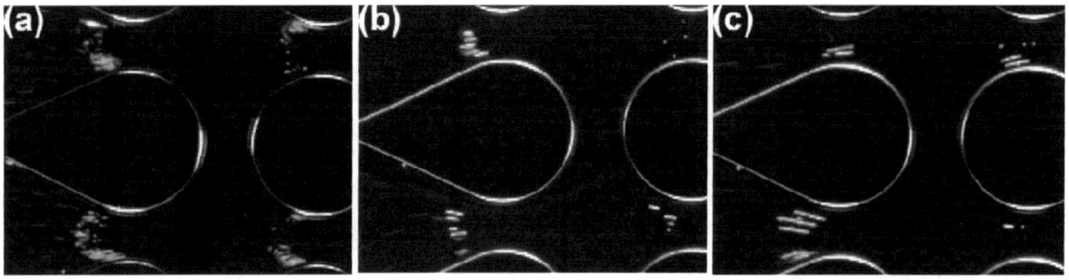

Figure 4. Shape of trapping regions as a function of particle concentration ratio ($C_{p,s}$:$C_{p,l}$) for binary systems having 100-nm and 1.14-µm-radius polystyrene particles. (a) System 2-1a with concentration ratio of 10:1 and ϕ = of 300 V. (b) System 2-2b with concentration ratio of 500:1 and $\phi = 350$ V. **(c)** System 2-3a with concentration ratio of 1000:1, and $\phi = 350$ V.

Similar results are shown in Figure 4c when the small-to-large ratio is increased to 1000:1. Large particles are immobilized along the flow pathlines, and trapping occurs in "*pearl-chain*" formations. An applied potential of only $\phi = 300$ V was needed to trap the large particles. These results demonstrate that the presence of the small particles in each system does not affect the dielectrophoretic immobilization of the large particles, but only the shape of the trapping regions.

Conclusions

The selectivity of an insulator-based dielectrophoretic (iDEP) microdevice was evaluated employing binary particle systems of different sizes. Experimental and mathematical modeling was carried out and the effect of different parameters over the microdevice selectivity was assessed. A microdevice made from glass, containing microchannels with arrays of insulating cylindrical structures and four different sizes of polystyrene particles were employed. Direct current electric potentials were used to immobilize particles with negative dielectrophoresis.

Experimental and modeling work allowed characterizing selective dielectrophoretic trapping of the large particles in each one of the binary systems employed as a function of particle sizes, particle concentration, concentration ratio and magnitude of the applied electric potential. It is expected that these results will expand on the knowledge of insulator-based dielectrophoresis, and have the potential to provide guidelines for the selection of operating conditions and design of iDEP microdevices.

Literature cited

[1] Whitesides, G. M., *Nature* 2006, *442,* 368-373.
[2] Pysher, M. D., Hayes, M. A., *Anal. Chem.* 2007, *79,* 4552-4557.
[3] Lapizco-Encinas, B. H., Rito-Palomares, M., *Electrophoresis* 2007, *28,* 4521-4538.
[4] Gascoyne, P. R. C., Vykoukal, J., *Electrophoresis* 2002, *23,* 1973-1983.
[5] Voldman, J., *Ann. Rev. Biomed. Eng.* 2006, *8,* 425-454.
[6] Hughes, M. P., *Electrophoresis* 2002, *23,* 2569-25582.
[7] Hughes, M. P., Nanoelectromechanics in Engineering and Biology, CRC Press, Boca Raton, FL, 2002.
[8] Kang, Y., Li, D., Kalams, S., Eid, J., *Biomed. Microdev.* 2008, *10,* 243-249.
[9] Cummings, E. B., Singh, A. K., *Anal. Chem.* 2003, *75,* 4724-4731.
[10] Davalos, R. V., McGraw, G. J., Wallow, T. I., Morales, A. M., Krafcik, K. L., Cummings, E. B., Simmons, B. A., *Anal. Bioanal. Chem.* 2008, *390,* 847-855.

[11] Sabounchi, P., Morales, A. M., Ponce, P., Lee, L. P., Simmons, B. A., Davalos, R., *Biomed. Microdev.* 2008, *10,* 661-670.
[12] Simmons, B. A., McGraw, G. J., Davalos, R. V., Fiechtner, G. J., Fintschenko, Y., Cummings, E. B., *MRS Bulletin* 2006, *31,* 120-124.
[13] Regtmeier, J., Duong, T. T., Eichhorn, R., Anselmetti, D., Ros, A., *Anal. Chem.* 2007, *79,* 3925-3932.
[14] Ying, L. M., White, S. S., Bruckbauer, A., Meadows, L., Korchev, Y. E., Klenerman, D., *Biophys. J.* 2004, *86,* 1018-1027.
[15] Clarke, R. W., White, S. S., Zhou, D., Ying, L., Klenerman, D., *Angew. Chem.* 2005, *44,* 3747-3750.
[16] Lapizco-Encinas, B. H., Ozuna-Chacón, S., Rito-Palomares, M., *J. Chromatogr. A* 2008, *1206,* 45-51.
[17] Lapizco-Encinas, B. H., Simmons, B. A., Cummings, E. B., Fintschenko, Y., *Anal. Chem.* 2004, *76,* 1571-1579.
[18] Lapizco-Encinas, B. H., Simmons, B. A., Cummings, E. B., Fintschenko, Y., *Electrophoresis* 2004, *25,* 1695-1704.
[19] Suehiro, J., Zhou, G. B., Imamura, M., Hara, M., *IEEE Trans. Ind. Appl.* 2003, *39,* 1514-1521.
[20] Ozuna-Chacón, S., Lapizco-Encinas, B. H., Rito-Palomares, M., Martínez-Chapa, S. O., Reyes-Betanzo, C., *Electrophoresis* 2008, *29,* 3115-3122.
[21] Martínez-López, J. I., Moncada-Hernández, H., Baylon-Cardiel, J. L., Martínez-Chapa, S. O., Rito-Palomares, M., Lapizco-Encinas, B. H., *Anal. Bioanal. Chem.* 2009, *394,* 293-302.
[22] Kwon, J.-S., Maeng, J.-S., Chun, M.-S., Song, S., *Microfluid. Nanofluid.* 2008, *5,* 23-31.
[23] Baylon-Cardiel, J. L., Lapizco-Encinas, B. H., Reyes-Betanzo, C., Chávez-Santoscoy, A. V., Martínez Chapa, S. O., *Lab Chip* 2009, *9,* 2896-2901.
[24] Pethig, R., Markx, G. H., *Trends Biotechnol.* 1997, *15,* 426-432.
[25] Ermolina, I., Morgan, H., *J. Colloid Interface Sci.* 2005, *285,* 419–428.

(372e) Size, Charge, and Affinity Fractionation of Hemolyzed Sera From a Neonatal Repository

Gary B. Smejkal[1], Camilia R. Martin[2], Steven Freedman[2] and Winston P. Kuo[1], (1)Harvard Catalyst, Laboratory for Innovative Translational Technologies, Harvard Medical School, Boston, MA, (2) Department of Neonatology, Beth Israel Deaconess Medical Center, Boston, MA

Hemolysis of serum or plasma can exclude very valuable samples from longitudinal studies, particular over time course, and risks the loss of valuable data. In human sera, the concentration of albumin is nearly ten billion times greater than that of cell-signaling proteins like the interleukins. The relative proportions of such rare, but extremely important biomarkers is further decreased in hemolyzed samples in which hemogloblin may nearly double the total protein concentration. Further, hemogloblin removal alone does not significantly decrease sample complexity. Sample prefractionation ideally partitions high abundance proteins like albumin, IgG, and hemoglobin into fractions separate from those containing the protein(s) of interest, potentially enabling enrichment and increasing the sensitivity of downstream microassays attempting to quantify these critical molecules. Samples from a neonatal depository are hoped to provide valuable information on pro-inflammatory response during the first days of life. However, such samples are extremely rare and frequently hemolyzed. To derive meaningful data from hemolyzed whole bloods and sera, fractionation schemes based of protein molecular mass, charge, or affinity were investigated as means of hemoglobin depletion with concomitant biomarker enrichment. Size separations were performed by sequential ultrafiltration with decreasing molecular weight cutoffs (MWCOs) to produce distinct size fractions, or by electrophoresis in the GELFREE System. Charge fractions were evaluated in the OFFGEL Fractionator or a Digital Protein Chip (dPC) capable of producing fractions of 0.05 pI unit intervals as well as pH specific column binding and elution. For affinity based separations, The ProteoMiner immobilized hexapeptide ligand library was used to isolate low abundance proteins without requiring prior hemoglobin depletion.

See more of this Session: Poster Session for the American Electrophoresis Society
See more of this Group/Topical: Topical 3: 2010 Annual Meeting of the American Electrophoresis Society (AES)

Spermatogenic Cells Manipulation Employing Dielectrophoresis

Elizabeth Rosales-Cruzaley, Perla A. Cota-Elizondo, Daniel Sánchez, and Blanca H. Lapizco-Encinas[*]
Email: blapizco@cinvestav.mx & *dpaulo@cinvestav.mx*
Microscale Bioseparations Laboratory and Biophysics of Ion Channels and Nanotechnology Laboratory, CINVESTAV-Monterrey, Apodaca NL, Mexico.

Introduction

Spermatozoa are produced in the testis during a process of differentiation known as spermatogenesis, this process involve a series of cell divisions, in which diploid spermatogonia undergo multiple rounds of proliferation (mitosis), differentiation (two meiotic divisions) and ultimately differentiate into mature haploid spermatozoa. A precise interplay between somatic cells (Leydig and Sertoli cells) and germ cells is the major signal that controls this tissue homeostasis in spermatogenesis. The spermatogenesis occurs inside the seminiferous tubules, where many cell types in various stages of development can be found, among them the whole sperm lineage (spermatogonias, primary and secondary spermatocytes, spermatids and sperms), that they differs so much in form, size and genetic content.

The spermatozoa are terminally differentiated cells, which mean that lacks of the machinery for protein synthesis. Therefore, all the proteins required for the sperm function of fertilize the egg, must be synthesized during spermatogenesis.

Many research studies are carried out with sperm and spermatogenic cells to understand the fertilization process in mammals. These range from morphology observations, motility assays, gene and protein expression to electrophysiological characterization to explore the role of ion channels in the different physiological processes, like motility, capacitation and acrosome reaction. These studies require mature sperm and spermatogenic cells obtained directly from testes. Therefore, there is an important need of techniques for fast purification and separation of the different cell types found in the testes. Traditionally, spermatogenic cells separation has been performed using the standard STA-PUT unit gravity sedimentation system using a linear BSA gradient.[1] This technique usually *i.e.,* separation gradients) requires large volumes of sample, is performed at cool temperatures and is time consuming and have low recovery. Some research studies require only a small sample and would benefit greatly from a technique capable of purifying a low volume of cells in a short time.

Microscale separation techniques are becoming increasingly more important in many biological applications. Miniaturization has attractive advantages such as shorter analysis time, higher resolution and sensitivity, lower requirement of sample and reagents.[2] Cell analysis is becoming one of the most important fields for lab-on-a-chip devices.[3,4] Manipulating, concentrating and sorting of cells employing electric fields offers great potential and flexibility.[4] Dielectrophoresis (DEP) is an electrokinetic transport mechanism where particles and cells can be concentrated and manipulated due to polarization effects produced by nonuniform electric fields. DEP is a nondestructive technique suitable for performing cells analysis. DEP-based microdevices have been employed for the differentiation of live from dead cells,[5,6] detection of cancer cells,[7-9] and simultaneous concentration and separation of mixtures of cells.[10] Traditionally, DEP has been applied employing arrays of microelectrodes, approach that has produced many important findings. However, electrode-based DEP has some drawbacks; one of the most important is that cells come in direct contact with the electrodes, leading to cell damage. An alternative is to employ insulator-based DEP (iDEP), a technique were nonuniform electric fields are created using an array of insulating structures across an electric field.[11] In this approach, electrodes are positioned remotely, reducing cell damage. Another important feature of iDEP is that electroosmotic flow can be used to pump fluid and cells across a microchannel or chamber, eliminating the need of a micropump, thus, simplifying the system.

This work presents the application of iDEP for the concentration and separation of mouse and rat spermatogenic cells. Samples of spermatogenic cells were introduced into a microchannel fabricated from polidimethylsiloxane, and by applying direct current electric fields it was possible to immobilize and concentrate the cells across an array of cylindrical insulating structures. There is a great potential for iDEP to be used for rapid sorting and separation of sperm cells for application where low volumes of sample are required.

Experimental

Sample preparation

Spermatogenic cells were obtained from testes of adult CD1 mice and/or Wistar rats. The testes were suspended in a Whitten-HEPES buffer NaCl 135 mM, KCl 5 mM, NaH_2PO4 0.5 mM, $CaCl_2$ 2 mM, HEPES 10 mM, dextrosa 10 mM used to ensure cell viability. The testes were decapsulated by cutting the tunica albuginea to expose the seminiferous tubules which were then mechanically disaggregated with tweezers and cut in smaller pieces using dissection scissors. Spermatogenic cells were released from the tubules by incubation in 2 mL of trypsin at a concentration of 0.25% and 2 mL of collagenase at a concentration of 0.001% for 5 minutes at 28°C. The resulting cell suspension was gently pipetted and were washed by centrifugation at 12,000 rpm for 5 minutes, resuspended in 5% glucose solution and then filtered with a cell strainer with a 40 µm opening (BD Falcon, Franklin Lakes, NJ). A volume of 500 µL was employed for labeling by adding 3 µL of syto 9 fluorescent dye (Invitrogen, Carlsbald California) ex/em 483/503 nm. Excess dye was eliminated by centrifugation of the cells for 5 minutes at 10,000 rpm and resuspending in 200 µL of glucose solution at 5% with pH= 7.3 and conductivity of 39 µS/cm.

Microdevice description

The microdevice employed was fabricated with standard soft lithography techniques used to pattern the microchannels that were then cast into poly(dimethylsiloxane) (PDMS) and sealed onto a glass slide. The microchannel employed in these experiments is 10.16 mm-long, 2 mm-wide, and 20 µm-deep. The posts are arranged in four rows and eight columns and have a diameter of 440 µm and separated 520 µm center-to-center. The first and last columns have "dove-tail" geometry to prevent particle agglomeration by collision with the posts and in this way preventing plugging of the system. A picture of the microdevices is presented in Figure 1a and a schematic representation of the microchannel used is shown in Figure 1b.

Equipment employed for dielectrophoresis experiments

A high voltage sequencer was used to apply DC electric potentials (model HVS448, LabSmith, Livermore, CA). The response of the cells was recorded in the form of videos and pictures employing an inverted epifluorescence video microscope for microfluidics, model SVM340 (LabSmith, Livermore, CA).

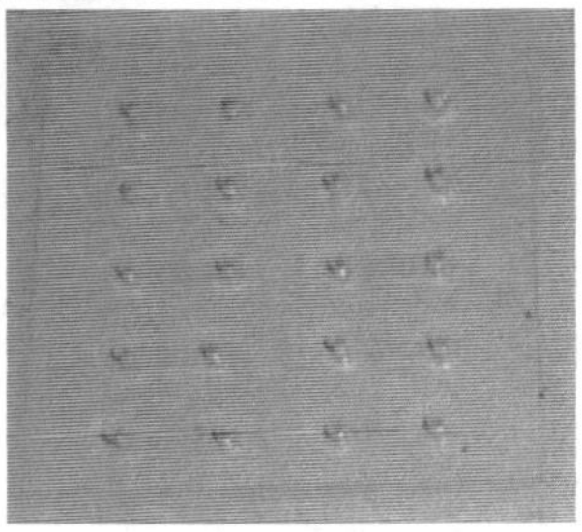

(a)

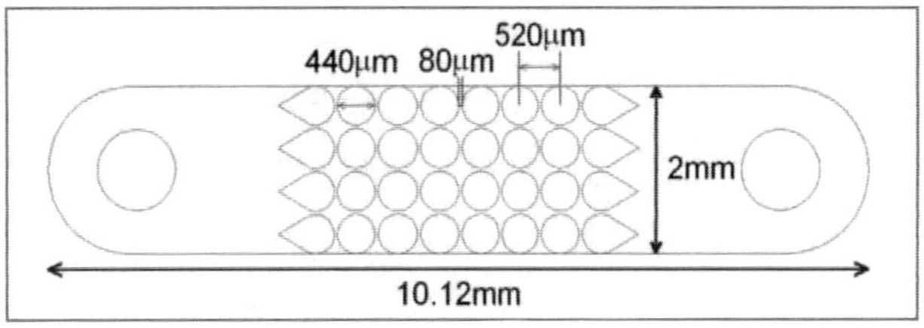

(b)

Figure2. (a) Picture of the microdevice employed. (b) Schematic representation of the microchannel employed for dielectrophoresis experiments.

Dielectrophoresis experimental procedure

In order to start experimentation, the microchannel was first cleaned with deionized water, and then filled with 5% glucose solution with pH = 7.3 and a conductivity of 39 μS/cm. Then, 10 μL of sample containing the labeled sperm cells was introduced at the inlet reservoirs. Platinum electrodes were then placed at the reservoirs and a DC electric potential was applied across the microchannel. The response of the cells was observed at different applied potentials and recorded employing the microscope camera.

Results and discussion

Cells were labeled with a fluorescent dye to improve visualization, Figure 2a shows a picture of the labeled spermatogenic cells where it is possible to observe the different stages of development; this is a sample of the cells employed for dielectrophoresis experimentation. Experiments were carried out by introducing a sample of the labeled cells into the inlet reservoirs and then applying a DC electric potential, and recording their response to the applied field in videos and pictures. It was observed that at low applied fields cell movement is dominated by electrokinetic flow and cells flow through the microchannel along streamlines,[12] as shown in Figure 2b obtained by applying an electric field of 200 V/cm. In order to obtain to cell immobilization higher electric fields have to be used, in order for the dielectrophoretic force to overcome electrokinetics. It was observed that by applying a field of 600 V/cm some cells were dielectrophoretically immobilized (Figure 2c); stronger dielectrophoretic responses were obtained by increasing the applied field to 1000 V/cm (Figure 2d). These results demonstrate that it is possible to manipulate cells employing iDEP, and are the first step towards the development of a fast dielectrophoretic-based method to concentrate and sort spermatogenic cells.

Conclusions

Microscale separation techniques are becoming a standard tool in many biological applications, especially those that require handling of cells. Dielectrophoresis, a nondestructive electrokinetic transport mechanism, is one of the most employed techniques for cells analysis and probing on the microscale.

The present work demonstrated that spermatogenic cells can be successfully manipulated and concentrated employing insulator-based dielectrophoresis. Spermatogenic cells were concentrated in an array of cylindrical insulating structures made from polidimethylsiloxane by applying direct current electric fields. Cell responses were observed as a function of the magnitude of the applied potential, where at low applied fields cells movement was dominated by electrokinetic flow, and dielectrophoretic immobilization occurred at higher applied fields.

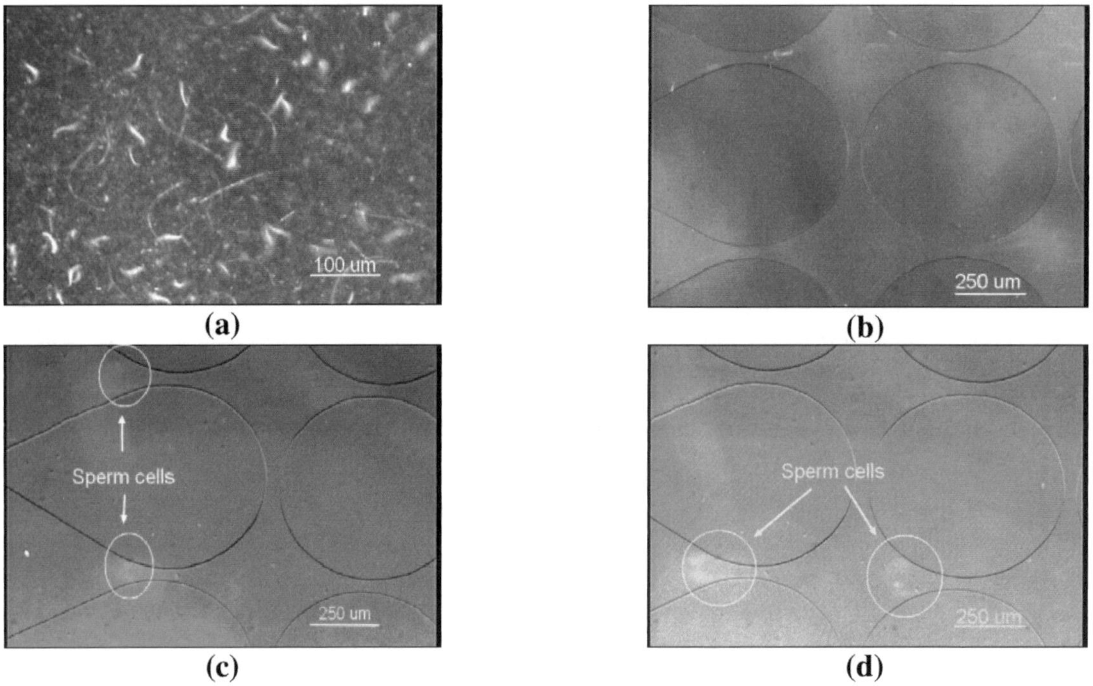

Figure 2. (a) Sample of the labeled spermatogenic and sperm cells. (b) Spermatogenic and sperm cells flowing across the microchannel when a potential of 200 V was applied. (c) Spermatogenic and sperm cells showing dielectrophoretic immobilization at a potential of 600 V. (d) Dielectrophoretic trapping of cells obtained at a potential of 1000 V.

The results presented here are one of the first studies of the application of a dielectrophoresis-based method for the manipulation of sperm cells. More work is required to fully explore the potential of this fast and powerful separation technique applied to sperm cells assessments. Future work includes determination of cell viability after dielectrophoretic manipulation. Additionally, separated and sorted cells can be used for different studies such as DNA and RNA extraction and physiological tests for the measurement of intracellular calcium.

Acknowledgements

The authors wish to thank Dr. Rafael Davalos and Hadi Shafiee from Virginia Tech for the fabrication of the microdevices. The authors acknowledge financial support provided by CINVESTAV-Monterrey.

References

1 Romrell L.J., Bellvé A.R., Fawcett D.W., (1976), "Separation of Mouse Spermatogenic Cells by Sedimentation Velocity: A Morphological Characterization," *Dev. Biol.* 49, pp.119-131.

2 Kang Y., Li D., Kalams S., Eid J., (2008), "DC-Dielectrophoretic separation of biological cells by size," *Biomed. Microdev.* 10(2), pp.243-249.
3 Minerick A.R., (2008), "The rapidly growing field of micro and nanotechnology to measure living cells," *AIChE J.* 54(9), pp.2230-2237.
4 Voldman J., (2006), "Electrical forces for microscale cell manipulation," *Ann. Rev. Biomed. Eng.* 8, pp.425-454.
5 Lapizco-Encinas B.H., Simmons B.A., Cummings E.B., Fintschenko Y., (2004), "Insulator-Based Dielectrophoresis for the Selective Concentration and Separation of Live Bacteria in Water," *Electrophoresis* 25, pp.1695-1704.
6 Li H., Bashir R., (2002), "Dielectrophoretic separation and manipulation of live and heat-treated cells of Listeria on microfabricated devices with interdigitated electrodes," *Sens. Actuator B-Chem.* 86(2-3), pp.215-221.
7 Gascoyne P.R.C., Wang X.B., Huang Y., Becker F.F., (1997), "Dielectrophoretic separation of cancer cells from blood," *IEEE Trans. Ind. Appl.* 33(3), pp.670-678.
8 Pohl H.A., Hawk I., (1966), "Separation of Living and Dead Cells by Dielectrophoresis," *Science* 152(3722), pp.647-649.
9 Gascoyne P.R.C., Noshari J., Anderson T.J., Becker F.F., (2009), "Isolation of Rare Cells from Cell Mixtures by Dielectrophoresis.," *Electrophoresis* 30(8), pp.1388-1398.
10 Moncada-Hernández H., Lapizco-Encinas B.H., (2010), "Simultaneous concentration and separation of microorganisms: insulator-based dielectrophoretic approach," *Anal. Bioanal. Chem.* 396(5), pp.1805-1816.
11 Cummings E.B., Singh A.K., (2003), "Dielectrophoresis in Microchips Containing Arrays of Insulating Posts: Theoretical and Experimental Results," *Anal. Chem.* 75(18), pp.4724-4731.
12 Martínez-López J.I., Moncada-Hernández H., Baylon-Cardiel J.L., Martínez-Chapa S.O., Rito-Palomares M., Lapizco-Encinas B.H., (2009), "Characterization of electrokinetic mobility of microparticles in order to improve dielectrophoretic concentration," *Anal. Bioanal. Chem.* 394(1), pp.293-302.

Particle manipulation in a multi-section insulator-based dielectrophoresis microdevice

Roberto C. Gallo-Villanueva[1] and Blanca H. Lapizco-Encinas[2,*]
Email: blapizco@cinvestav.mx

[1] BioMEMS Reseach Chair, Tecnológico de Monterrey, Monterrey NL, Mexico.
[3] Microscale Bioseparations Laboratory, CINVESTAV-Monterrey, Apodaca NL, Mexico.

Introduction

Particle separation and concentration has become very important as an analytical method. Yet performing such operations is not simple due to similarities in physical and chemical properties of the analytes on complex samples. There are several techniques with the aim of purifying samples but there is still a wide area of opportunity in terms of resolution and efficiency. Driven by advances in electronics, miniaturization offers an alternative with minimal usage of substances and samples, decrease of wastes, shorter process times, greater sensitivity and portability. Nowadays fabrication techniques allow for novelty designs to improve separation technologies.

One of such technologies is dielectrophoresis (DEP) which is the electrokinetic transport mechanism of particles due to polarization effects in the presence of non-uniform electric fields.[1] The most common approach to apply DEP is with AC electric fields employing arrays of microelectrodes.[2] However, manufacture of microdevices with electrode arrays can be expensive and operation affected by electrode degradation.[3] An alternative to produce non-uniformity of the electric field is to straddle insulator structures between two electrodes, i.e., insulator-based dielectrophoresis (iDEP).

Despite the novelty of the technique, there have been important microdevice designs to separate mixtures of particles according to their dielectrophoretic response. Advances in microdevice fabrication open the possibility to include diverse non-uniformity of the electric field within different parts of a single channel, and therefore separation of complex mixtures can be achieved. In 2005, Barret, et al.,[4] introduced a device with a three-dimensional ridge-like channel which changed in depth, thus creating different zones for particle trapping by means of iDEP. Later, in 2007, Thwar, et al.,[5] presented a dynamic system where dielectrophoretic traps are manipulated by the size oil menisci. The channel use two different sets of menisci which could be shaped in different sizes and therefore having two dielectrophoretic traps to separate samples. In the same year, Pysher and Hayes[6] developed a saw-tooth microchannel which tooth dimensions change along the channel. In this way the width of the channel decreased in the direction of the flow increasing the gradient of the electric field, and therefore the DEP force. In this way they could separate live and dead samples of Bacillus subtilis, Escherichia coli and Staphylococcus epidermis in different regions of the same microdevice.

In the present study, insulating cylindrical posts are used to produce the non-uniform electric field and dielectrophoretically trap polystyrene microspheres. The microdevice employed was manufactured in glass and has diverse post diameters that create different regions with their own DEP magnitudes. In this way, manipulating the electric field, the microspheres can be trapped in the desired section of the channel where negative dielectrophoretic force overcomes electroosmotic flow.

Theory of dielectrophoresis

As described in Pohl,[1] 1951, dielectrophoresis is an electrokinetic transport mechanism driven by polarization effects on particles when subjected to non-uniform electric fields. As mentioned earlier, such field with regions of high and low electric field intensity can be produced by the inclusion of insulator structures between two electrodes, which is referred to as insulator-based dielectrophoresis. The dielectrophoretic force exerted on a spherical particle can be defined as:

$$F_{DEP} = 2\pi\varepsilon_m r_p^{\,3}\,\mathrm{Re}\left(f_{CM}\right)\,\nabla E^2 \qquad (1)$$

where ε_m is the permittivity of the suspending medium, r_p is the particle radius, ∇E is the gradient of the electric field, and $\mathrm{Re}(f_{CM})$ is the real part of the Clausius-Mossotti (CM) factor which can be approximated to the real conductivities when low frequency electric fields are used:

$$f_{CM} = \left[\frac{\sigma_p - \sigma_m}{\sigma_p + 2\sigma_m} \right] \quad (2)$$

where σ_p and σ_m are the conductivities of the particle and suspending medium respectively.

There are other electrokinetic forces present when using DC electric fields. Electrokinetic force (EK) is the superposition of electrophoresis (EP) and electroosmotic flow (EOF), and for a microchannel with a negative surface charge, such as glass, the electrokinetic velocity (v_{EK}) can be defined as follows, considering the net direction of the flow as positive[7]:

$$v_{EK} = \mu_{EK} E = (\mu_{EO} - \mu_{EP}) E \quad (3)$$

$$v_{EO} = \mu_{EO} E \quad (4)$$

$$v_{EP} = \mu_{EP} E \quad (5)$$

where μ_{EO} and μ_{EP} are the electroosmotic and electrophoretic mobilities, and v_{EO} and v_{EP} are the electroosmotic and electrophoretic velocities, respectively.

To achieve trapping of particles, the dielectrophoretic velocity must overcome that of electrokinetics[7]. Since v_{EK} is linearly related to the field while v_{DEP} has a relationship of second order with the applied field, it is possible to achieve this condition under high electric fields:

$$v_{DEP} = -\mu_{DEP} \nabla E^2 \quad (6)$$

where v_{DEP} and μ_{DEP} are the dielectrophoretic velocity and mobility, respectively.

Experimental section

Microdevice employed

The microdevice was fabricated from two D263 borosilicate glass substrates, where the top substrate contains reservoirs and the bottom substrate holds the channels. The channel used has three sections with cylindrical insulator posts of different diameters for each one, as shown in Figure 1. Each section has a side-channel to recover separated and concentrated sample, nevertheless, for this work, all side-channel reservoirs were sealed and therefore not used. The first set of posts have a diameter of 175 μm and are arranged 408.3 μm center-to-center. The middle section has posts with 275 μm in diameter and are separated 408.3 μm center-to-center. The insulators in the third segment have a diameter of 335 μm and are arranged 388.3 μm center-to-center. The first row of posts for each segment has a "dove-tailed" geometry to prevent particles to collide to the posts and plugging the system.

Equipment

Direct current electric fields were applied using a 3000 V high voltage sequencer, model HVS448 (LabSmith, Livermore, CA, USA) by employing platinum-wire electrodes (Omega, Stamford, CT, USA) at each of the reservoirs. Dielectrophoretic behavior of the microspheres was visualized with the use of an inverted microscope model Axiovert 200 (Carl Zeiss, Göttingen, Germany), which has an integrated camera, and recorded as video utilizing the Microscope Software AxioVision LE (Carl Zeiss, Göttingen, Germany). A 10x microscope objective was used for all experiments. The microscope is equipped with a UV lamp that was used for excitation of the sample. Both devices require the use of a personal computer.

Sample preparation and DEP experiments

Buffer was prepared with bidistilled water and adjusted to pH 6 and conductivity of 150 μS/cm by adding NaOH and KH_2PO_4. Polystyrene microspheres of 1 μm in diameter (Invitrogen, Carlsbad, CA, USA), excitation/emission 505/515 nm, were resuspended in buffer to a concentration of $1x10^8$ spheres/mL.

The microchannel was filled with buffer solution and sample was introduced in the inlet reservoir. Electric fields were applied using the high voltage sequencer by placing the electrodes in the inlet and outlet reservoirs.

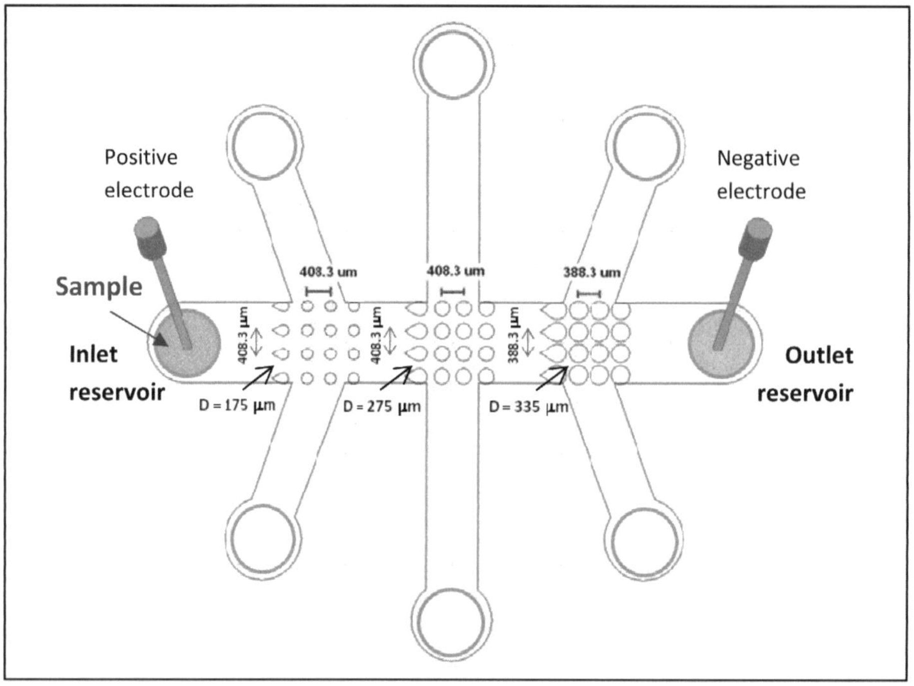

Figure 1. Schematic representation of the microdevice employed. Insulator posts are arranged in three sections with different cylindrical insulator posts in each one (D = diameter of posts). Sample is loaded in the inlet reservoir and electric fields are applied between the inlet and outlet reservoirs. Lateral reservoirs were sealed and not used.

Results and discussion

From equation (1) it can be inferred that a higher gradient of the electric field is reflected on a higher magnitude of the DEP force. Each section of the microdevice has a different effect on the electric field exerted in the system. This means that when a specific electric field is applied between the inlet and outlet reservoirs, its gradient will be different for each post size. The biggest insulators will have the greatest gradient of the electric field and therefore the highest value for the DEP force. Thus, a smaller value for the electric field will be needed to overcome EK and achieve particle trapping. In order to get a similar effect on the dielectrophoretic behavior for example on the second section of the device as on the third, a higher electric field will be needed.

Figure 2 shows results obtained with the device described when different electric fields were applied using green-fluorescent microspheres of 1 µm in diameter. As it can be seen, an electric field of 1500 V/cm is not enough for DEP to overcome EK and particles flow dominated by EOF using insulator posts of 275 µm in diameter, but by increasing the electric field to 2500 V/cm strong DEP trapping can be observed (Figure 2a). By employing insulators with 335 μm-diameter, weak DEP trapping is presented with an electric field of 800 V/cm and the DEP is clearly increased with a field of 1200 V/cm (Figure 2b). In the section with insulator posts of 175 μm in diameter no DEP trapping was observed during the experiments with any electric fields applied (results not shown). These results were obtained with a suspending medium of pH 6 and conductivity of 150 μS/cm.

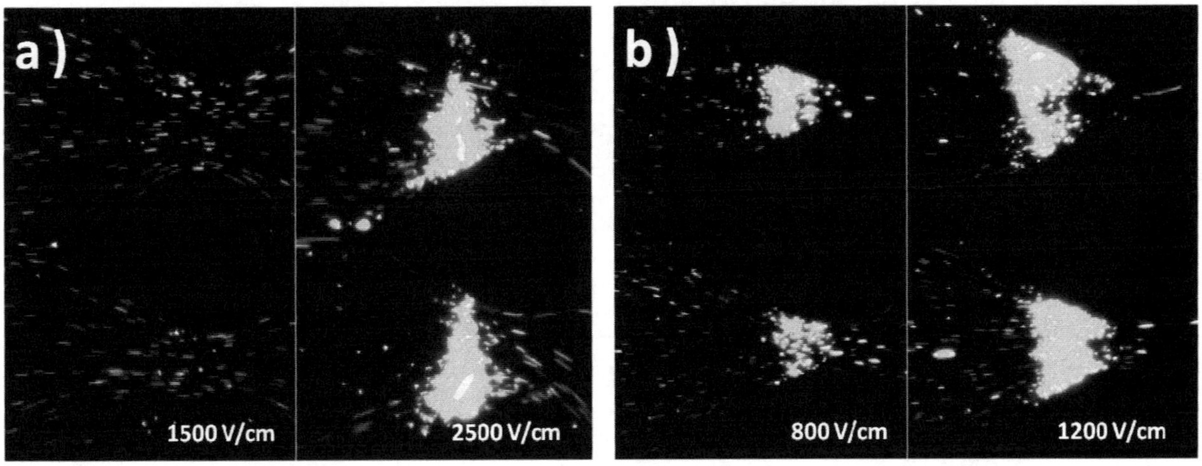

Figure 2. Dielectrophoretic response of polystyrene microspheres of 1 μm with a suspending medium of pH 6 and conductivity of 150 μS/cm. a) Using posts with a diameter of 275 μm and an electric field of 1500 V/cm particles are dominated by EOF, and DEP trapping is achieved by increasing the field to 2500 V/cm; b) with the insulators of 335 μm in diameter particles are weakly trapped by the DEP force and strong trapping is achieved with an electric field of 1200 V/cm.

With these buffer conditions (pH 6 and conductivity of 150 μS/cm) negative DEP was observed. This is represented by repulsion of the particles from the regions of higher field intensity (narrow part between posts), i.e., the microspheres are accumulated just before the regions.

Conclusions

A microdevice with different insulator sizes was used to trap and concentrate polystyrene spheres. The place where particles wanted to be concentrated could be manipulated with the electric field applied between the reservoirs. Negative DEP was observed for the particles since their conductivity is lower than that of the buffer used, which can be seen in the region where they were concentrated, that is just before the region of high electric field intensity (narrow part between posts). The magnitude of the applied electric field could be manipulated to trap or release the particles in desired regions. The repulsion from the regions of higher field intensity, negative DEP trapping, could be also controlled with the electric field.

This work shows the potential of the technique to separate complex mixtures of particles in a microchannel applying the same operation conditions to the system (buffer pH and conductivity and electric field).

References

1. Pohl, H.A. 1951. The motion and precipitation of suspensoids in divergent electric fields. Journal of Applied Physics. Vol. 22: 869.

2. Washizu, M., Kurosawa, O. 1990. Electrostatic manipulation of DNA in microfabricated structures. IEEE Transactions on Industry applications Appl. 26, 1165-1172.

3. Lapizco-Encinas, B.H., Simmons, B.A., Cummings, E.B., Fintschenko, Y. 2004. Dielectrophoretic Concentration and Separation of Live and Dead Bacteria in an Array of Insulators. Analytical Chemistry. 76, 1571-1579.

4. Barrett, L.M., Skulan, A.J., Singh, A.K., Cummings, E.B., Fiechtner, G.J. 2005. Dielectrophoretic Manipulation of Particles and Cells Using Insulating Ridges in Faceted Prism Microchannels. Analytical Chemistry. 77, 6798-6804.

5. Thwar, P.K., Linderman, J.J., Burns, M.A. 2007. Continuous Electrodeless Dielectrophoretic Separation in a Circular Channel. Journal of Physics: Conferences Series. 34, 527-532.

6. Pysher, M.D., Hayes, M.A. 2007. Electrodeless direct current dielectrophoresis using reconfigurable field-shaping oil barriers. Electrophoresis 28, 4572-4581.

7. Martínez-López, J.I., Moncada-Hernández, H., Baylon-Cardiel, J.L., Martínez-Chapa, S.O., Rito-Palomares, M., Lapizco-Encinas, B.H. 2009. Characterization of electrokinetic mobility of microparticles in order to improve dielectrophoretic concentration. Analytical and Bioanalytical Chemistry. 394(1):293-302.

(372h) Electro-Poiseuille Flow Modeling in Annular Geometry

Jeffery W. Thompson, Seth Wynne, Holly Stretz, Mario Oyanader and Pedro Arce, Chemical Engineering, Tennessee Technological University, Cookeville, TN

The effect of geometry on the separation of ions in electrolytic solution is studied. The rectangular and cylindrical cases have already been derived as the result of previous work. For the case of steady-state flow in the annular region between two fixed, charged, concentric cylinders, the profiles of concentration, electric potential, and total mixture velocity have been obtained assuming the electrolytic fluid to be dilute in ion concentration. A decoupling technique is employed to handle the separation of the hydrodynamic problem from the electrostatics problem. For the electrostatics, the Debye-Huckel approximation is used, however a numerical method for improving this linearization of the Poisson-Boltzmann equation is discussed and shown to provide excellent convergence characteristics. Finally, the species continuity equation is stated (in terms of the electro-migration flux) and solved.

See more of this Session: Poster Session for the American Electrophoresis Society
See more of this Group/Topical: Topical 3: 2010 Annual Meeting of the American Electrophoresis Society (AES)

(372i) Using a Linear Operator Method Approach in Assessing Dynamic Behaviors in Electrophoresis Separations

Jennifer Pascal and Pedro Arce, Chemical Engineering, Tennessee Technological University, Cookeville, TN

The technique of electrophoresis is relevant in a variety of chemical engineering applications, including soil remediation, separation of biomacromolecules, and drug delivery, just to name few. In many of these applications, the dynamic behaviors play an important role in controlling performance of the device used or the implementation needed for such applications. Therefore, it is crucial to have a better understanding of the behavior of electrophoresis from the dynamics point of view in order to help improve or optimize systems where electrophoresis is an important aspect of their operation.

This contribution will present an operator theoretic formulation relevant for an analysis of the dynamics of, for example capillary electrophoresis, free-flow-electrophoresis and other similar systems. Our aim here is rather modest and it will be focused on determining essential steps useful to explain the role of key parameters associated with the systems. For example, we will offer a characterization of potential dynamic behaviors in term of spectral-based information associated with the transport operator of the system. Furthermore, an attempt will be made to extract information from the spectrum that can help to a priori determine favorable conditions for an improved separation. Typical dynamic behaviors will be illustrated for a more concrete appreciation of the benefits of using operator-theoretic methods in applied field operations.

See more of this Session: Poster Session for the American Electrophoresis Society
See more of this Group/Topical: Topical 3: 2010 Annual Meeting of the American Electrophoresis Society (AES)

(372j) Temperature Distribution in Electrochromatography with An Oscillatory Transverse Electric Field

Wei Yuan[1], David R. Nielsen[2] and Yan Sun[1], (1)Chemical Engineering, Tianjin University, Tianjin, China, (2)Department of Chemical Engineering, Arizona State University, Tempe, AZ

A three-compartment column of ion-exchange electrochromatography with an oscillatory electric field perpendicular to mobile-phase flow driven by pressure (pIEEC) was verified for its high throughput purification of proteins[1]. Electroosmotic flow was proved to be major contributor to its mass transfer acceleration[2]. By modeling work, protein distribution in adsorbents was found to present excursion along the electric field direction. The electro-kinetic convection in porous particle was predicted as exponentially decreased function of protein adsorption amount[3]. So that, at the beginning of the pIEEC, intraparticle convection caused by the electric field contributed more to the enhancement of dynamic binding capacity. For utilization of the new design, an experimental system for in-column temperature measurements was constructed, and the dynamic processes of the in-column temperature in electrochromatography were examined. With the experimental system, the effect of electric current strength and mobile-phase ionic strength on the in-column temperature was investigated. Then, a heat transfer model for the pIEEC was established for the dynamic process modeling. The mathematical model concerns force and free convection heat transfer, as well as Darcy modified wall effect in packed bed column. It was confirmed that the model was in good agreement with the measurements. By the model calculations, we could also obtain the in-column voltage distribution, effective voltage applied to the central compartment and the efficiency of energy consumption. Moreover, the model was used for scale-up analysis and for investigation of temperature distribution under various conditions.

1. Journal of Separation Science, 2006, 29, 2383; 2. Separation and Purification Technology, 2009, 68, 109-113 3. Electrophoresis, 2010, 31, 944-951

See more of this Session: Poster Session for the American Electrophoresis Society
See more of this Group/Topical: Topical 3: 2010 Annual Meeting of the American Electrophoresis Society (AES)

(372k) Monitoring Algae Species in Bio-Fuel Production by CE-SSCP

Alice C. Jernigan[1], Lauren Woods[1], Robert Beitle[1], Jamie Hestekin[2] and Christa Hestekin[1], (1) Chemical Engineering, University of Arkansas, Fayetteville, AR, (2)Chemical Engneering, University of Arkansas, Fayetteville, AR

With its ability to use carbon dioxide from the atmosphere as a carbon source, and to clean nitrates and phosphorous from waste water, algae has the potential to be a significantly eco-friendly bio-fuel source. One of the problems faced in using algae as a bio-fuel is identifying and characterizing the best species for the purpose, and keeping them as the dominate species. The identification of algal communities can be accomplished by isolating DNA from water samples. These DNA samples can then be characterized by a technique called single-strand conformational polymorphism (SSCP) analysis of the 18S gene. The 18S gene is highly conserved among eukaryotes, but has unique small variable regions that can be used to differentiate between species. These regions can be amplified by polymerase chain reaction (PCR) giving small fragments of the gene sequence. SSCP, a method that is normally used to detect mutations, is the heat denaturing of these fragments resulting in single strand DNA. The refolding of the single strand DNA will form different secondary structures (conformations) depending on the nucleotide sequence. These different conformations can be separated due to their different electrophoretic mobility, even if they are the same size. The increased speed and automation of capillary electrophoresis (CE)-SSCP will allow for more rapid and reliable results allowing a quick and inexpensive method for monitoring the different algae species.

See more of this Session: Poster Session for the American Electrophoresis Society
See more of this Group/Topical: Topical 3: 2010 Annual Meeting of the American Electrophoresis Society (AES)

(415a) Imaging Amperometry: Single Particle Experiments and Theory

Christopher L. Wirth, Paul J. Sides and Dennis C. Prieve, Department of Chemical Engineering and Center for Complex Fluids Engineering, Carnegie Mellon University, Pittsburgh, PA

The ability to measure local electrochemical current density is important to a number of applications including the high throughput evaluation of electrocatalysts and corrosion studies. We have developed a method called *Imaging Amperometry*[1,2] that measures current density via an optical signal arising from a colloidal particle near a working electrode. The particle's height is a function of the intensity and direction of local current density, so by precisely measuring height to nanometer resolution, we infer local current density on the scale of the particle's diameter (~5 μ m). Measurements are conducted using a new instrument, called the Electrochemical Total Internal Reflection Microscope (ETIRM), that combines techniques traditionally found in colloid science (TIRM) and electrochemistry (cyclic voltammetry). TIRM is used to monitor scattered light from a particle, as light intensity is exponentially related to the particle's height. A 3-electrode electrochemical measurement is conducted along with TIRM. We demonstrate this technique using a continuous Indium Tin-Oxide (ITO) thin film electrode, a discontinuous ITO thin film electrode, and a continuous (but electrochemically patchwork) ITO/gold thin film electrode. In all cases, we are able to measure local electrochemical activity by monitoring the scattered light intensity from the particle. In addition to these experimental studies, theory is presented to predict current density from the optical measurement.

1. Patent pending 2. P.J. Sides, C.L. Wirth, D.C. Prieve, *An imaging ammeter for electrochemical measurements*. Electrochemical and Solid State Letters, Accepted April 2010

See more of this Session: Nanoscale Electrokinetics
See more of this Group/Topical: Topical 3: 2010 Annual Meeting of the American Electrophoresis Society (AES)

(415b) Multiple Measurements On Single DNA Molecules in Nanopore Sensors

Jason Yi-Heng Sen and Rohit Karnik, MIT, Cambridge, MA

There is a significant gap in technology for rapid, label-free characterization of single molecules and particles in the submicrometer size range. Advances in nanotechnology have resulted in nanoscale Coulter counters and nanopores that can size single molecules and particles using transient current changes during translocation through the nanopore. The sensing ability of nanopores is often limited by the quick transit times of molecules through the nanopore that results in poor signal-to-noise ratio. To address this issue, we fabricated a nanopore device and implemented feedback control to recapture DNA back into the nanopore, yielding multiple measurements and enabling statistical averaging on the same molecule. With the implementation, we demonstrated that different DNA molecules that have slight differences in magnitude of current changes could be differentiated in our nanofluidic system. This approach may lead to nanopore devices with enhanced sensitivity for analysis of single molecules and particles.

See more of this Session: Nanoscale Electrokinetics
See more of this Group/Topical: Topical 3: 2010 Annual Meeting of the American Electrophoresis Society (AES)

(415c) Nanopore Sensing and Characterization of Hepatitis B Viral Capsids

Kaimeng Zhou[1], Lichun Li[2], Zhenning Tan[2], Adam Zlotnick[2] and Stephen C. Jacobson[1], (1) Department of Chemistry, Indiana University, Bloomington, IN, (2)Department of Molecular and Cellular Biochemistry, Indiana University, Bloomington, IN

Shrinking the lateral dimensions of micromachined channels to nanometer length scales provides unique opportunities for sensing and separation applications. Some aspects of microchannel transport transfer directly to operation of smaller nanoscale channels, but nanofluidic systems can be significantly influenced by phenomena such as double layer overlap, surface charge, ion current rectification, diffusion, and entropic forces, which are either insignificant or absent in larger microchannels. For example, funnel-shaped nanochannels rectify ion current, and nanofunnels with smaller taper angles exhibit higher current rectification. These devices are also used to sense transiting particles, e.g., viruses, and to characterize particle properties, e.g., surface charge and payload. We will present characterization of Hepatitis B Viral (HBV) capsids that translocate through a single conical nanopore, which is chemically etched in a poly(ethylene terephthalate) (PET) membrane after being tracked by a heavy ion. The nanopore is chemically modified to minimize the electroosmotic flow inside the pore as well as charge-charge interactions between the capsid and pore wall. The HBV capsids are electrophoretically driven through the nanopore, and their presence is sensed by a change in the ionic current as the capsids transit through the nanopore. We demonstrate that discrimination of HBV capsids with different sizes (a T=3 capsid is ~31 nm in diameter and a T=4 capsid is ~37 nm in diameter) is feasible based on the difference in the change in ionic current. These results illustrate the potential of these nanopore devices for sensitive size characterization of biomolecules or particles in real time.

See more of this Session: Nanoscale Electrokinetics
See more of this Group/Topical: Topical 3: 2010 Annual Meeting of the American Electrophoresis Society (AES)

(415d) Applications of Nanoparticle Organic Hybrid Materials in DNA Separation

Henry W. Lau and Lynden A. Archer, Department of Chemical and Biomolecular Engineering, Cornell University, Ithaca, NY

Comprising of silica nanoparticle cores with a covalently-linked corona of polyethylene glycol oligomers, Nanoscale Organic Hybrid Materials (NOHMs) are a recently-developed class of materials which exhibit liquid-like properties in the absence of a solvent. This talk focuses on using NOHMs to enhance the electrophoretic separation of DNA. In particular, we report on the incorporation of silica nanoparticle hybrids into a polyacrylamide gel sieving matrix, where the mobility of eluting DNA fragments are affected in a size-dependent manner. We discuss our findings in context of the steric and electrostatic properties of the NOHMs, and we present our efforts in developing a theoretical framework for DNA electrophoresis that accounts for these properties.

See more of this Session: Nanoscale Electrokinetics
See more of this Group/Topical: Topical 3: 2010 Annual Meeting of the American Electrophoresis Society (AES)

(415e) Fluid-Flow Enhanced Electrokinetic Sample Pre-Concentration in a Voltage Gated Nanochannel

Aditya S. Khair and Todd M. Squires, Chemical Engineering, University of California at Santa Barbara, Santa Barbara, CA

Recent years have witnessed much interest in electrically-driven, or electrokinetic, ion-selective transport through nano-fluidic channels for e.g. sample pre-concentration, sensing, ionic current rectification, energy conversion, and single-molecule studies. A common feature among this diverse set of applications is that ion-selectivity occurs as a result of overlapping nano-meter sized electrical double layers surrounding charged nanochannel walls. In this talk, we present a theoretical analysis of ion-selective transport and sample pre-concentration in a nanochannel containing a tunable "gate" electrode. An imposed axial electric field drives ionic transport along the nanochannel; ion-selectivity of the channel is controlled by tuning the (zeta) potential of the gate. Importantly, ions are enriched (pre-concentrated) and depleted at opposite sides of the gate, and we consider the possibility of enhancing the degree of enrichment using pressure-driven fluid flow along the nanochannel. Indeed, a simple scaling analysis suggests that flow can lead to an exponential increase in enrichment. However, a detailed asymptotic and numerical treatment reveals that this enhancement can be lost at large flow rates, for which we provide a physically-based explanation.

See more of this Session: Nanoscale Electrokinetics
See more of this Group/Topical: Topical 3: 2010 Annual Meeting of the American Electrophoresis Society (AES)

(415f) Dielectrophoresis and AC-Field Induced Encapsulate Release of Micelles in Aqueous Suspensions

Victoria Froude and Yingxi Elaine Zhu, Chemical and Biomolecular Engineering, University of Notre Dame, Notre Dame, IN

Recently, lipid based biocolloids, such as micelles and liposomes, have been of increasing interest as drug delivery systems for controlled release, specific cell targeting, or medical diagnostics via advanced electrokinetic and microfluidic techniques. In this work, we examine dielectrophoresis (DEP) of complex bio-nanocolloids, where nanoparticles can be effectively and rapidly manipulated and assembled in suspensions by the use of an AC-electric field. The dynamic response of micelles, of 3-5 nm in diameter, to applied AC-electric fields is examined through the use of fluorescence correlation spectroscopy (FCS) at a single-molecule resolution. We focus on the AC-field induced transport of sodium dodecyl sulfate (SDS) micelles tagged with various fluorescent encapsulates in aqueous media. The AC-electric field induced concentration changes and DEP mobility are accurately determined over a broad range of AC-field frequencies (5 kHz – 20 MHz) at amplitude of 10 V_{pp}. We observe a strong AC-frequency dependence of micelle concentration between two microelectrodes, from which the DEP crossover frequency is determined. The crossover frequency, or the point at which the nanocolloid DEP response switches between positive and negative DEP, is observed for micelles and nanocolloids of 10 nm for the first time. Surprisingly, we also observe an AC-field induced dissociation of the micelle structure and a resulting release of fluorescent encapsulates at a characteristic low AC-field frequency of ~1-10 kHz, where the dissociation has been found to be dependent on the surface charge of the interior encapsulate. Manipulation and induced release of complex nanostructures could have potential applications for controlled drug release and lab-on-a-chip technologies.

See more of this Session: Nanoscale Electrokinetics
See more of this Group/Topical: Topical 3: 2010 Annual Meeting of the American Electrophoresis Society (AES)

Microparticle Manipulation Employing Low Frequency AC Electric Fields with Insulator-Based Dielectrophoresis

Javier L. Baylon-Cardiel,[1] Nadia M. Jesús-Pérez,[1] Ana V. Chávez-Santoscoy,[1] and Blanca H. Lapizco-Encinas[2,]*
Email: blapizco@cinvestav.mx

[1] BioMEMS Reseach Chair, Tecnológico de Monterrey, Monterrey NL, Mexico.
[2] Microscale Bioseparations Laboratory, CINVESTAV-Monterrey, Apodaca NL, Mexico.

Abstract

Low frequency insulator-based dielectrophoresis is a novel technique that allows for highly controlled manipulation of particles. By varying the shape of an AC signal applied across a microchannel containing an array of insulating cylindrical structures it was possible to concentrate and immobilize microparticles in bands; and then, move the bands of particles to a different location. Mathematical modeling was performed to analyze the distribution of the electric field and electric field gradient as a function of the shape of the AC applied potential, employing frequencies in the 1 Hz range. Experimental results demonstrated that this novel dielectrophoretic mode allowed highly modulated dielectrophoretic manipulation.

Introduction

The increasing interest on microsystems for applications on diverse areas demands the development of techniques for particle manipulation at the microscale. The use of electric fields to achieve the manipulation is a useful approach, since its application can produce the motion of both the particles of interest and their suspending medium, through the different electrical forces that are present. DEP is the motion of particles by the application of a non-uniform electric field, and it depends on the effect of the induced charge on a particle. This phenomenon was first described by Pohl,[1] and it has been proposed as a useful technique for the manipulation of particles[2] achieving the separation and concentration of biological particles, such as proteins,[3-5] DNA,[6-8] and bacteria,[9-11].

Dielectrophoresis can be carried out with either alternating current (AC) or direct current (DC) electric fields and can be applied to both charged and neutral particles. Generally, AC fields are employed with arrays of microelectrodes. Electrode-based DEP has been successfully applied for the manipulation of many particles, but devices are complex and expensive to fabricate.[12] A simpler and less expensive approach to carry out DEP is insulator-based DEP (iDEP). In this novel technique non-uniform electric fields are created by employing insulators, that act as obstacles that bend the electric field, leading to regions of higher and lower magnitude of the electric field. The majority of the studies with iDEP have been carried out using DC electric fields.[10,13,14] Very few reports on the implementation of iDEP with AC electric fields are available.[4,13,15,16]

In this work, we present the manipulation of polystyrene particles in a microchannel with cylindrical insulating posts, employing an AC electric field. Full groups or bands of trapped particles were manipulated depending on the form of the AC signal employed (sinusoidal, half sinusoidal) using frequencies below the range of 1 Hz, which are significantly lower when compared to the majority of the reports on AC-DEP studies. This novel dielectrophoretic mode introduces another parameter to modulate dielectrophoretic mobilization of particles and/or groups of particles.

Theory of dielectrophoresis

For a time variant applied electric potential, the motion induced on a particle due to the electrical force exerted on it can be expressed in terms of the time dependant electrokinetic velocity

(EK) as:

$$\vec{u}_{EK} = \mu_{EK}\vec{E}_s f(t) \quad (1)$$

where μ_{EK} is the electrokinetic mobility, $\vec{E}_s$ is the spatially non uniform electric field, and $f(t)$ is a periodic function of time. The dielectrophoretic force on a spherical particle of radius R exerted by an electric field with magnitude E is defined as

$$\vec{F}_{DEP} = 2\pi R^3 \varepsilon_m f_{CM} \nabla E^2 \quad (2)$$

where ε_m is the permittivity of the suspending medium and f_{CM} is the Classius-Mossotti (CM) factor, which accounts for particle polarizability. The CM factor is dependent on the frequency of the applied electric potential; however, for low frequencies, the CM factor can be expressed in terms of the conductivities of the particle (σ_p) and the suspending medium as (σ_m):

$$f_{CM} = \left[\frac{\sigma_p - \sigma_m}{\sigma_p + 2\sigma_m}\right] f_{CM} = \left[\frac{\sigma_p - \sigma_m}{\sigma_p + 2\sigma_m}\right] \quad (3)$$

This is the same condition as for an applied DC potential.[17] The contribution of the DEP force to the motion of the particle can be expressed in terms of the dielectrophoretic velocity:

$$\vec{u}_{DEP} = -\mu_{DEP}\nabla E^2 \quad (4)$$

where μ_{DEP} is the dielectrophoretic mobility. For a spherical particle of radius R, the mobility μ_{DEP} is defined as:

$$\mu_{dep} = \frac{R^2 \varepsilon_m f_{CM}}{6\eta} \quad (5)$$

where η is the suspending medium viscosity. For a time dependant electric field with magnitude $E = E_s f(t)$, the DEP velocity can be written as:

$$\vec{u}_{DEP} = -\mu_{DEP}\nabla E_s^2 [f(t)]^2 \quad (6)$$

In order to achieve DEP immobilization of a particle, its contribution to the particle motion must overcome other transport mechanisms, such as diffusion and EK motion. In this study, it is assumed that the main contributions to the flux of particles $\vec{j}$ along the microchannel come from EK and DEP, as

$$\vec{j} \approx \vec{u}_{EK} + \vec{u}_{DEP} \quad (7)$$

It can be assumed that particle immobilization is achieved in the regions were the condition $\vec{j} \cdot \vec{E} = 0$ is satisfied. In terms of the EK and DEP mobilities, this condition can be expressed as:

$$(\mu_{EK}\vec{E} - \mu_{DEP}\nabla E^2) \cdot \vec{E} = 0 \quad (8)$$

Moreover, in terms of a time dependant electric field, the condition for trapping can be expressed as function of time as:

$$\left(\frac{\mu_{DEP}\nabla E_s^2}{\mu_{EK} E_s^2} \cdot \vec{E}_s\right) f(t) > 1 \quad (9)$$

Mathematical model and experimental section

Mathematical model

The geometry of the iDEP device was captured with *COMSOL Multiphysics* software, the microchannel considered was 10.16-mm long, 10-μm deep and 2-mm wide containing an array of 32 cylindrical insulating posts, distributed in four rows of eight posts each. The posts had a diameter of

200 μm, arranged 250 μm center-to-center (Figure 1a). For the model, 1 μm particles were considered. The conductivity of the suspending medium was assumed to be much larger than the conductivity of the particles and so the CM factor was approximated to the value of -0.5. A triangular mesh was built in order to solve the Laplace equation using the *Lagrange element* implemented in the software. Finite element analysis was applied in a triangular mesh consisting of 600 elements and 1800 nodal points. The electric potential V was approximated by a polynomial of low order at each mesh point. From the numerical solution it is possible to obtain values for the parameters describing the time dependant motion of the particles, such as the electrical field $\vec{E}$ and the gradient of E^2 for different times along a period of the AC signal.

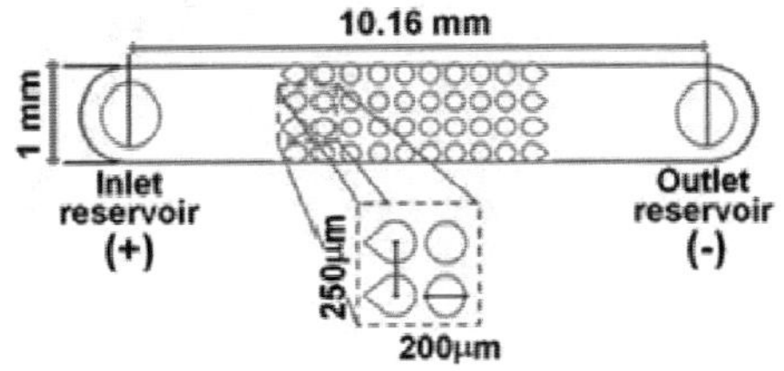

Figure 1. Schematic representation of the microchannel employed for iDEP simulations and experiments.

iDEP experiments

Experiments were carried out in a glass composed by two wafers of Schott 263D glass. The microchannel and the cylindrical posts were wet etched on the bottom wafer with photolithography and wet etchings techniques; reservoirs were drilled on top wafer, and the microdevice was completed by thermal bonding of the two glass wafers (Fig 1).

Carboxylated-modified fluorescent green-yellow 1-μm-diameter microspheres (Invitrogen, Carlsbad, CA, USA), ex/em 505/515 nm, were used in a concentration of 7.639 x10^{10} microspheres ml^{-1}. The suspending medium was distillated water with pH of 6 and conductivity of 2.45 μS cm^{-1}. Image and video capturing was done employing an inverted microfluidic microscope SVM340 with an integrated video camera (LabSmith, Livermore, CA, USA). A high voltage power supply HVS448 (LabSmith, Livermore, CA, USA) was used to generate the electric fields with DC and AC signals. Prior to each experiment, the microchannel was cleaned and filled with the suspending medium. A sample of 40 μl of the microspheres was introduced at the inlet reservoir. Care was taken to eliminate pressure-driven flow, and then electrodes were placed at the inlet and outlet reservoirs in order to apply a DC electric field prior the AC signal.

Results and discussion

Mathematical modeling and different set of experiments were carried out in order to analyze the particle behavior with two different shapes of AC signals

Sinusoidal signal

Simulation

By considering a Sinusoidal signal with a frequency of 1 Hz and amplitude of 750 V (Figure 2), a set of simulations was performed in order to predict the electric field and electric field gradient distribution across the array of insulating posts. Figures 3a and 3b show the electric field distribution and Figures 3c and 3d show the gradient of E^2 when the signal shown in Fig. 2 is applied. It is important to note that the distribution of the electric field presents two variations, as it can be seen in Figures 3. One variation comes from the spatially non-uniform distribution of the electric field because the insulating posts act as obstacles for the electric field. The other variation comes from the time dependence of the field on a periodic signal. In order to obtain dielectrophoretic immobilization of particles, DEP has to overcome EK (Eq. 9), this only occurs at higher applied potential. In Figure 2 can be observed in red the values of the applied potential and time where EK dominates and in blue the values where DEP dominates.

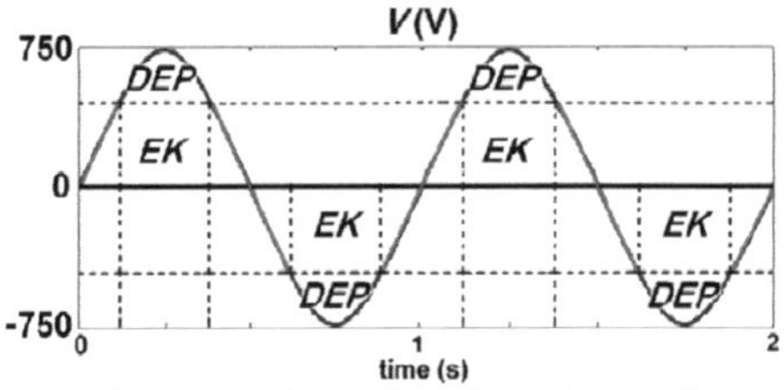

Figure 2. Schematic representation of the sinusoidal signal with amplitude 750 V and period T= 1 s (f = 1 Hz).

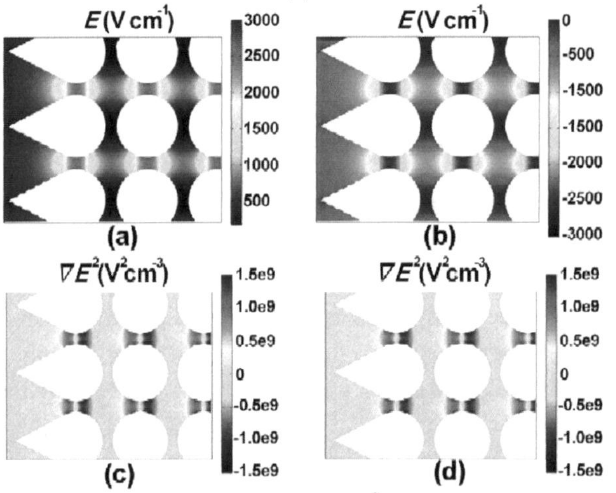

Figure 3. Distribution of the electric field E and the gradient of E^2 for different times: (a) E at $t=\pi/2$, (b) E at $t=3\pi/2$, (c) gradient of E^2 at $t=\pi/2$ and (d) gradient of E^2 at $t=3\pi/2$.

The most positive value of E (3000 V cm^{-1}) is shown in Figure 3a at time $t=\pi/2$ sec, a particle experiencing EK motion at this time would move from right to left. As time passes, the field decreases until it reaches it most negative value (-3000 V cm^{-1}) at time $t=3\pi/2$ sec (Fig 3b), a particle under EK motion would move from left to right. Overall, the average value of the displacement of the particle due to EK over a period is zero. In contrast, as shown in Figures 3c and 3d, the magnitude and distribution of the gradient of E^2 are alike at times t = $\pi/2$ sec and t = $3\pi/2$ sec. The maximum and minimum values of the gradient of E^2 are achieved at time t =$\pi/2$ (Fig. 3c) and as time passes, the magnitude decreases to its minimum value at time t = π sec. After this time, it increases again to its maximum value at time t = $3\pi/2$ sec (Fig. 3d). Thus, the gradient of E^2 only takes positives values. When DEP overcomes EK motion, particles are repelled away from the regions of higher gradient since under these conditions polystyrene particles exhibit negative dielectrophoretic behavior. Particles become immobilized. *i.e.*, form bans of trapped particles, in the regions where the condition in Equation (9) is satisfied.

Experiments

Experimentally, it was observed that a frequency of 0.4 Hz was needed to achieve successful particle manipulation at an applied potential of 750 V, these results are presented in Figure 4. Figure 4a show that articles under EK motion move from the left to the right, until reaching the electric field magnitude maximum. At this time, the particle movement is dominated by DEP and particle trapping is achieved in the left side of the posts, forming a band of particles (Fig. 4b). Then, the cycle repeats, the field magnitude and EK dominates over DEP; and particles move from left to right. When half the period has passed and the electric field is negative, the EK flow changes its direction, going from right to left, as shown in Figure 4c. Then, at the most negative DEP again dominates EK, and particles are trapped again (Fig. 4d). This results show how the shape of the applied signal can tune the particle movement from EK to DEP.

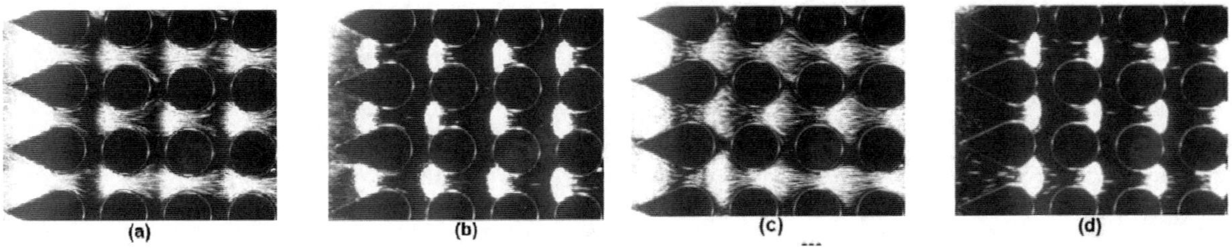

Figure 4. Response of the trapping bands to the sinusoidal signal at 0.4 Hz for different times. (a) Particles moving from left to right due to EK; (b) Particles are trapped due to DEP; (c) Particles moving from right to left due to EK; (d) Particles are trapped due to DEP, again.

Half-sinusoidal signal

Simulation

A modification of the sinusoidal signal, presenting only the positive half of the total signal over the period was tested to analyze particle behavior. A representation of the signal is included in Figure 5. The signal was obtained by using a sinusoidal signal with amplitude of 750 V and a period of 4 s, and applying only on intervals of 2 s. Electric potential and time values here EK dominates particle motion are shown in red and DEP-dominated values are shown in blue (Fig. 5). These conditions allow transporting band or group of particles in unified manner from one column of posts to the next. At low magnitudes of the applied signal, EK dominates and particles move from left to right, then at the potential maxima, DEP dominated and particles are immobilized. As the potential decreases to zero and increases from zero EK dominates again, the group/band of particles move again from left to right due to EK, then as the potential reaches the maxima, particles are immobilized again. This cycle repeats, where the particles go from moving to trapping, going downstream the array of insulating structures.

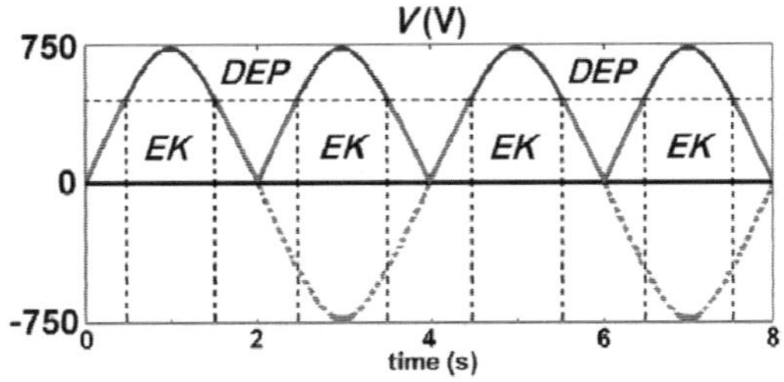

Figure 5. Schematic representation of the half sinusoidal signal with amplitude 750 V and period T= 4 s (f = 0.25 Hz).

Experiments

Experiments employing a half sinusoidal signal were carried out and the results are included in Figure 6. Highly controlled manipulation of particles was obtained, where bands of trapped particles were formed as shown in Figure 6a. When the signal is applied, as EK dominates the flow of particles (shown in red in Fig. 5), the bands of immobilized particles disappear, releasing the particles that then move along the electric field lines. Particles transport is greatly dependant on the amplitude and period of the employed AC signal. In this case, for the employed amplitude of 750 V the optimal period for the complete sinusoidal signal was 4 s. It would be expected that for a lower amplitude, the value of the period would have to increase in order to move particles from one column of insulating posts to the next. This is because particles would experience a smaller EK velocity and, in order to move along the distance between posts, they need to travel for a longer interval of time under EK flow. When DEP starts to dominate the flow of particles (shown in blue in Fig. 5), bands of immobilized particles are reformed in the adjacent posts, as shown in Figure 6b. For a time equal to 10 periods of the half sinusoidal signal, the transport of the bands of trapped particles is obtained, as presented from Figure 6c to 6e, by looking at the "***thicker***" bands, which are in the first column in Fig. 6a, then second column in Figure 6b, third column in Figure 6c, and so on. These results demonstrate how particles can be released from the post array in a modulated and controlled manner.

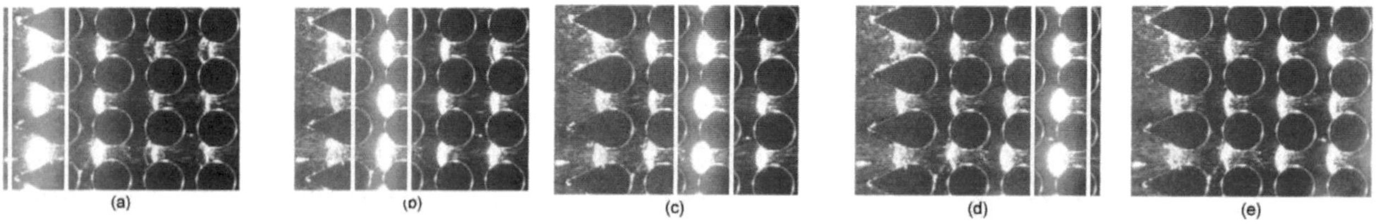

Figure 6. Transport of trapped particles employing the half sinusoidal signal, it is possible to identify the "*thicker*" bands going downstream. (a) Thicker bands of trapped particles are located at the beginning of the array (b) The bands are split and particles move from left to right and the bands are reformed, now the thicker bands are at the second column of posts. (c) Thicker bands are reformed at the third column. (d) Thicker bands are reformed at the fourth column. (e) Image showing that the thicker band of particles have exited the array of posts.

Conclusions

The successful manipulation of bands of trapped particles employing two different shapes of low frequency AC voltages in a microchannel with an array of cylindrical insulating structures has been demonstrated. Experimental and modeling work was performed for a sinusoidal and a half sinusoidal signal. Simulations are shown to be in agreement with the experimental results that show how bands or groups of particles could be transported across the post array, moving between columns of adjacent insulating posts and obtaining the reallocation of the trapping band, demonstrating highly controlled manipulation and release of groups of particles.

Low frequency AC-DEP (~ 1 Hz) has great potential for highly controlled manipulation of individual particles or groups of particles, since the shape of the AC voltage represents another parameter used to direct and modulate particle motion. This novel low frequency dielectrophoretic mode could be used to improve the separation and fractionation of mixtures of particles, a common challenge in many biotechnological applications.

References

1 Pohl H.A., (1951), "The Motion and Precipitation of Suspensoids in Divergent Electric Fields," *J. Appl. Phys.* 22(7), pp.869-871.

2 Castellanos A., Ramos A., González A., Green N.G., Morgan H., (2003), "Electrohydrodynamics and dielectrophoresis in microsystems: scaling laws," *J. Phys. D: Appl. Phys.* 36, pp.2584-2597.

3 Lapizco-Encinas B.H., Ozuna-Chacón S., Rito-Palomares M., (2008), "Protein Manipulation with Insulator-Based Dielectrophoresis and DC Electric Fields," *J. Chromatogr. A* 1206(1), pp.45-51.
4 Clarke R.W., White S.S., Zhou D., Ying L., Klenerman D., (2005), "Trapping of proteins under physiologycal conditions in a nanopipette," *Angew. Chem.* 44, pp.3747-3750.
5 Zheng L.F., Brody J.P., Burke P.J., (2004), "Electronic manipulation of DNA, proteins, and nanoparticles for potential circuit assembly," *Biosens. Bioelectron.* 20(3), pp.606-619.
6 Yokokawa R., Manta Y., Namura M., Takizawa Y., Le N.C.H., Sugiyama S., (2010), "Individual evaluation of DEP, EP and AC-EOF effects on [lambda]DNA molecules in a DNA concentrator," *Sensors and Actuators B: Chemical* 143(2), pp.769-775.
7 Gallo-Villanueva R.C., Rodríguez-López C.E., Díaz-de-la-Garza R.I., Reyes-Betanzo C., Lapizco-Encinas B.H., (2009), "DNA manipulation by means of insulator-based dielectrophoresis employing direct current electric fields," *Electrophoresis* 30(24), pp.4195-4205.
8 Bakewell D.J., Morgan H., (2006), "Dielectrophoresis of DNA: time-and frecuency-dependent collections on microelectrodes," *IEEE Trans. Nanobios.* 5(1), pp.1-8.
9 Moncada-Hernández H., Lapizco-Encinas B.H., (2010), "Simultaneous concentration and separation of microorganisms: insulator-based dielectrophoretic approach," *Anal. Bioanal. Chem.* 396(5), pp.1805-1816.
10 Cho Y.-K., Kim S., Lee K., Park C., Lee J.-G., Ko C., (2009), "Bacteria concentration using a membrane type insulator-based dielectrophoresis in a plastic chip," *Electrophoresis* 30(18), pp.3153-3159.
11 Suehiro J., Hamada R., Noutomi D., Shutou M., Hara M., (2003), "Selective detection of viable bacteria using dielectrophoretic impedance measurement method," *J. Electrostatics* 57(2), pp.157-168.
12 Cummings E.B., Singh A.K., (2003), "Dielectrophoresis in Microchips Containing Arrays of Insulating Posts: Theoretical and Experimental Results," *Anal. Chem.* 75(18), pp.4724-4731.
13 Chou C.F., Tegenfeldt J.O., Bakajin O., Chan S.S., Cox E.C., Darnton N., Duke T., Austin R.H., (2002), "Electrodeless dielectrophoresis of single- and double-stranded DNA," *Biophys. J.* 83(4), pp.2170-2179.
14 Suehiro J., Zhou G.B., Imamura M., Hara M., (2003), "Dielectrophoretic filter for separation and recovery of biological cells in water," *IEEE Trans. Ind. Appl.* 39(5), pp.1514-1521.
15 Ying L.M., White S.S., Bruckbauer A., Meadows L., Korchev Y.E., Klenerman D., (2004), "Frequency and voltage dependence of the dielectrophoretic trapping of short lengths of DNA and dCTP in a nanopipette," *Biophys. J.* 86(2), pp.1018-1027.
16 Baylon-Cardiel J.L., Lapizco-Encinas B.H., Reyes-Betanzo C., Chávez-Santoscoy A.V., Martínez Chapa S.O., (2009), "Prediction of trapping zones in an Insulator-based dielectrophoretic device," *Lab Chip* 9(20), pp.2896-2901.
17 Ermolina I., Morgan H., (2005), "The electrokinetic properties of latex particles: comparison of electrophoresis and dielectrophoresis," *J. Colloid Interface Sci.* 285, pp.419–428.

(415h) Predicting Electroosmotic Volumetric Flowrates In Porous Media: Role of Capillary Geometries In the Description of Media Morphology

Jennifer Pascal, Mario Oyanader and Pedro Arce, Chemical Engineering, Tennessee Technological University, Cookeville, TN

Electrokinetic-based methods are used in a variety of applications including drug delivery and separation of biomolecules, among others. Many of these applications feature a fibrous or a porous medium that can be modeled by using capillary bundle models to predict the behavior of the electroosmotic flow within the particular system. The role of geometry in predicting volumetric flowrates in porous media is investigated by modeling the electroosmotic flow in idealized capillaries of rectangular, cylindrical, and annular geometries. This is achieved by the coupling of electrostatics and continuum hydrodynamics to obtain analytical expressions that govern the electrokinetically - driven volumetric flow within these idealized capillary geometries. A previous study developed a model to compare the cylindrical and annular capillary geometries by utilizing two methods that compare the areas of the two geometries. (Wu and Papadopoulos, 2000). The methods used in this previous work will also be used in the present contribution to compare the volumetric flowrates in the cylindrical and annular capillaries with a rectangular capillary. Illustrative results will be presented to aid in the understanding of the influence of the various geometrical and electrostatic parameters that arise from the analysis of these volumetric flowrates. It was found that the magnitudes of the volumetric flowrates in the cylindrical and annular capillary geometries are larger than those of the rectangular capillary geometry.

See more of this Session: Nanoscale Electrokinetics
See more of this Group/Topical: Topical 3: 2010 Annual Meeting of the American Electrophoresis Society (AES)

(462a) Simulation of DNA Conformation Pre-Conditioning for Electrophoretic Stretching in a Microcontraction

Chih-Chen Hsieh, Chemical Engineering Department, National Taiwan University, Taipei, Taiwan and **Tsung-Hsien Lin**, National Taiwan University, Taipei, Taiwan

We use Brownian dynamics-finite element method (BD-FEM) to examine new approaches to improve DNA stretching by electrokinetic flow in a microfluidic device. The device is based on the design of Kim and Doyle[1] that stretches DNA by passing them through a hyperbolic contraction. To further enhance DNA stretching, we propose two strategies that pre-condition DNA before they enter the contraction. For the first approach, we pre-stretch DNA in the direction perpendicular to the funnel axis with a cross-slot-like geometry. The partially stretched chains are then turned to align with the axial of the funnel, and experience the second stretching. As a result, DNA chains adapt more extended configurations before going into the funnel, and therefore achieve a higher degree of extension. For the second approach, we pre-condition DNA conformation using an oscillating extensional flow that has been shown to effectively reducing the population of folded DNA at ideal condition. However, this approach shows insignificant effect on DNA stretching. Detailed comparison for the evolution of DNA configuration will be analyzed to provide the guidance for further optimization of our design.

References [1]. Kim, J. M. and P. S. Doyle (2007). "Design and numerical simulation of a DNA electrophoretic stretching device." Lab on a Chip 7(2): 213-225.

See more of this Session: DNA Analysis in Microfluidic and Nanofluidic Devices
See more of this Group/Topical: Topical 3: 2010 Annual Meeting of the American Electrophoresis Society (AES)

(462b) DNA Translocation through Nanopores Under Time-Varying Electric Fields: A Brownian Dynamics Study

Christopher M. Edmonds, School of Bioengineering, Georgia Institute of Technology, Atlanta, GA, Yeny C. Hudiono, Chemical and Biological Engineering, University of Colorado, Boulder, CO, Amir A. Ahmadi, Chemical and Biomolecular Engineering, Georgia Institute of Technology, Atlanta, GA, Peter Hesketh, Mechanical Engineering, Georgia Institute of Technology, Atlanta, GA and Sankar Nair, School of Chemical & Biomolecular Engineering, Georgia Institute of Technology, Atlanta, GA

The single-molecule analysis of biopolymers such as DNA and proteins has a wide range of potential applications in medicine and biotechnology. In particular, devices for high-speed and inexpensive analysis (e.g., sequencing) of biopolymers offer a route towards overcoming the limitations of conventional techniques like electrophoresis. One of the promising single-molecule methods is based on the use of nanopore devices, which are individual porous channels of diameters 1-5 nm and lengths 5-25 nm. Because DNA is negatively charged under usual solution conditions in water, it can be translocated through the nanopore with the aid of an applied electric field. By measuring the translocation time for DNA through the nanopore (usually in the range of ~ 1-10 μ s/base) and the modulations in ionic current through the pore caused by the presence of DNA, many characteristics of a DNA chain can be deduced such as the length and the sequence of base pairs.

Among the different types of nanopore devices being used in translocation time measurements, solid state (e.g., silicon nitride) nanopores, have had good success. Nanopore devices are currently operated using DC electric fields and cannot discriminate between biopolymers that differ only slightly in length. In previous work [1] we proposed, using a simple mathematical model, that the application of optimized electric fields would dramatically increase the resolution of the device in successfully discriminating biopolymers that differ only slightly in length. The aim of the present work is to develop insights into the translocation of DNA through nanopores under the application of time-varying electric fields, and to investigate the possibility of improved resolution of nanopore devices by this method.

We will first discuss the development of an accurate Brownian Dynamics model that captures all the essential dynamics associated with DNA translocation through a silicon nitride pore. DNA is modeled as a freely jointed chain using a bead-spring model, with the flexibility to increase the number of DNA monomers per bead and to use two different spring force laws (Fraenkel and Warner). Both the excluded volume effect and the interactions between DNA and the silicon nitride nanopore are modeled with a Lennard-Jones potential using parameters obtained from measured data. To accurately model the electrostatic effects in the system, we compute the potential due to the applied voltage and the charges associated with the atoms of the nanopore by Poisson's equation. The Langevin approach is used for integrating the equations of motion of the DNA chain, leading to the incorporation of stochastic thermal fluctuations into the model.

We then discuss the application of this model to study the transport of DNA in silicon nitride nanopores of diameter 1-5 nm and thickness 5-10 nm under the influence of both direct and time-varying electric fields. We investigate the effects of chain length, composition, and electrolyte conditions upon the translocation properties including the average translocation time and the statistical variance in translocation time due to thermal fluctuation. Using our detail computational method, we rigorously examine our earlier proposition that the use of time-varying electric fields (e.g., sinusoidal and square wave forms) substantially improve the resolution of DNA size analysis using nanopores. In particular, there is a "resonance effect" that occurs when the frequency of the applied electric field matches the characteristic time scale of the translocation process. We conclude by discussing the implications of our findings for design and operation of nanopore sensors for biomolecular analysis application.

References

[1]. Bhattacharya S., Nair S., Chatterjee A., An Accurate DNA Sensing and Diagnosis Methodology Using Fabricated Silicon Nanopores. IEEE Transactions on Circuits and Systems I: Regular Papers. 2006; 53(11): 2377-2383.

See more of this Session: DNA Analysis in Microfluidic and Nanofluidic Devices
See more of this Group/Topical: Topical 3: 2010 Annual Meeting of the American Electrophoresis Society (AES)

(462c) Molecular Simulation of Polyelectrolye Conformational Dynamics Under An AC Electric Field

Hongjun Liu, Yingxin Zhu and Edward Maginn, Chemical and Biomolecular Engineering, University of Notre Dame, Notre Dame, IN

We use a coarse-grained molecular dynamics method to study the behavior of a flexible polyelectrolyte (PE) chain in an explicit salt solution with varied valence under an AC electric field. Simulations in the absence of electric field and under DC electric field are used to determine the critical field strength and intrinsic chain relaxation frequency. Our results show that the PE chain breathes with applied AC frequency and becomes dynamically stretched, only when the applied field strength exceeds the critical field strength and the applied AC frequency is comparable to or less than the intrinsic relaxation frequency of the PE chain.

See more of this Session: DNA Analysis in Microfluidic and Nanofluidic Devices
See more of this Group/Topical: Topical 3: 2010 Annual Meeting of the American Electrophoresis Society (AES)

(462d) Pressure-Induced Flow and Resultant Orientation of Polyacrylamide Chains Impacts Peak Width in DNA Sequencing Separatoins in Microchannel Electrophoresis

Thomas Niedringhaus[1], Daniel G. Hert[2], Christopher P. Fredlake[2] and Annelise Barron[3], (1)Chemical Engineering, Stanford University, Stanford, CA, (2)Department of Chemical and Biological Engineering, Northwestern University, Evanston, IL, (3)Bioengineering, Stanford University, Stanford, CA

For years, microchannel electrophoresis has provided high-resolution separations of DNA, which in turn has lead to highly accurate sequencing data. Initially, microchannel electrophoresis was performed on a capillary array electrophoresis (CAE) platform. As the need to increase throughput has risen, much work has been done to transition from the CAE platform to a microchip electrophoresis platform. While microchip electrophoresis offers faster separation times, this technology has yet to offer the consistent long-reads of CAE. Previous work in our lab has shown differences in separation performance between CAE and microchip electrophoresis for the same replaceable polymer network; namely the separation efficiency greatly decreases in microchips. To better understand DNA migration in microchip, we looked at the apparent dispersion coefficient of DNA for a number of different linear polyacrylamides and poly(N,N-dimethylacrylamides) polymer networks to determine the extent of band broadening. In this study, the physical properties of linear polyacrylamide networks were found to influence band broadening in a time-dependent fashion. Ultimately, understanding the relationship between the physical properties of the polymer network and band broadening will result in increased read lengths on the microchip platform.

See more of this Session: DNA Analysis in Microfluidic and Nanofluidic Devices
See more of this Group/Topical: Topical 3: 2010 Annual Meeting of the American Electrophoresis Society (AES)

(462e) Electrophoretic Concentration of DNA at Nanoporous Polymer Membranes for Separations and Diagnostics

Robert J. Meagher and Anup K. Singh, Sandia National Laboratories, Livermore, CA

We report on the use of thin (~30 micron) photopatterned polymer membranes for on-line preconcentration of single- or double-stranded DNA samples prior to electrophoretic analysis. Shaped UV laser light is used to quickly (~10 seconds) polymerize a highly crosslinked polyacrylamide plug. By applying an electric field across the membrane, DNA from a dilute sample can be concentrated into a narrow zone (<100 micron wide) at the outside edge of the membrane. The field at the membrane can then be reversed, allowing the narrow plug to be cleanly injected into a separation channel filled with a sieving polymer for analysis. Concentration factors >100 are possible, increasing the sensitivity of analysis for dilute samples. We have fabricated both neutral membranes (purely size-based exclusion) as well as anionic membranes (size and charge exclusion), and characterized the rate of preconcentration as well as the efficiency of injection from both types of membrane, for DNA, ranging from a 20 base ssDNA oligonucleotide to >14 kbp dsDNA. We have also investigated the effects of concentration polarization on device performance for the charged membrane.

Advantages of the membrane preconcentration approach include the simplicity of device fabrication and operation, and the generic (non-sequence specific) nature of DNA capture, which is useful for complex or poorly characterized samples where a specific capture sequence is not present. The membrane preconcentration approach is well suited to simple single-level etch glass chips, with no need for patterned electrodes, integrated heaters, valves, or other elements requiring more complex chip fabrication. Additionally, the ability to concentrate multiple charged analytes into a narrow zone enables a variety of assay functionalities, including enzyme-based and hybridization-based analyses.

See more of this Session: DNA Analysis in Microfluidic and Nanofluidic Devices
See more of this Group/Topical: Topical 3: 2010 Annual Meeting of the American Electrophoresis Society (AES)

(462f) Long DNA Separation Using a Sparse Micropost Array

Jia Ou, Samuel J. Carpenter and **Kevin D. Dorfman**, Department of Chemical Engineering and Materials Science, University of Minnesota - Twin Cities, Minneapolis, MN

Long DNA are difficult to separate using traditional gel electrophoresis. We will present separation data obtained in a microfluidic device containing a sparse post array. The posts were made in hexagonal array of 1 μm diameter in 3 μm spacing using photolithography and deep trench etching. DNAs of λ XHo1 digest, Hind III digest, λ DNA-Mono cut mix, 2λ and T4 were mixed with λ DNA and tested in the device. The separations were measured with electropherograms recorded by a photomultiplier tube and a computer controlled fluorescent microscope. Under an electric field of 10 V/cm, the mobilities of different species were normalized by the λ DNA mobility in the mixtures and compared across the size range from 10 kbp to 165.5 kbp. DNA mobility depends linearly on molecular weight in this regime, allowing us to calibrate the device. We demonstrate the robustness of the device by using the calibration function to identify the peaks in a separation of the λ DNA-Mono cut mix.

See more of this Session: DNA Analysis in Microfluidic and Nanofluidic Devices
See more of this Group/Topical: Topical 3: 2010 Annual Meeting of the American Electrophoresis Society (AES)

(462g) DNA Separation and Sequencing at a Stretch

Jiamin Wu[1], Shuangliang Zhao[2], Lizeng Gao[1], Jianzhong Wu[2] and Di Gao[1], (1)Chemical and Petroleum Engineering, University of Pittsburgh, Pittsburgh, PA, (2)Department of Chemical and Environmental Engineering, University of California, Riverside, Riverside, CA

We are developing a novel DNA sequencing technology based on tethering DNA strands to a solid surface through precise end-hybridization and sequentially pulling DNA strands off the surface under an electric field. This new DNA fractionation method is expected to be much more efficient than the electrophoresis-based technologies for separation of DNA strands in different lengths produced from the Sanger reaction. Previously, we have been able to use this method to separate long DNA molecules by length. For example, we have demonstrated efficient separation of lambda dsDNA (48,502 bp) from human genomic dsDNA (> 100,000 bp). However, separation of short ssDNA fragments, which is required for DNA sequencing, by this method has been challenging, because an extremely large electric field is needed to load the short ssDNA strand with a large enough stretching force that can pull the ssDNA off the surface. This large electric field may cause various problems such as Joule heating and side electrochemical reactions in the buffer solution. Besides, the solvent-induced fluctuation force on DNA fragment becomes chain length dependent. Here we present two approaches that we have developed to overcome those challenges, through which short ssDNA strands (<100 bases) are efficiently separated. The first is based on applying a potential drop across a microchannel, which effectively reduces the current to less than several microamps and the temperature change to less than 1 K during the entire separation process. The second is based on pulling the ssDNA off a noble metal surface, which uses the Gouy-Chapman-Stern screening layer to generate a strong electric field for pulling the ssDNA. By using either approach, short ssDNA stands of 60, 80, and 90 bases are efficiently separated. General mechanism for the separation of short ssDNA fragments is demonstrated by considering the fluctuation forces on the basis of Brownian dynamics simulation.

See more of this Session: DNA Analysis in Microfluidic and Nanofluidic Devices
See more of this Group/Topical: Topical 3: 2010 Annual Meeting of the American Electrophoresis Society (AES)

(462h) Exploiting the Entropic Trapping Regime to Enhance Separation Performance in Microchip Gel Electrophoresis

Nan Shi and Victor M. Ugaz, Artie McFerrin Department of Chemical Engineering, Texas A&M University, College Station, TX

Electrophoretic DNA transport through nanoporous hydrogels is characterized by an entropic trapping (ET) mechanism when the DNA size is close to the gel pore size. This mode of transport can yield highly desirable transport properties (e.g., enhanced size dependences of mobility and diffusion for improved separations). But hydrogels have largely been ignored in efforts to understand and exploit ET effects, where the focus has instead centered on simpler idealized "nanofilter" architectures based on arrays of alternating deep wells and narrow slits (i.e., describable by only two length scales). We have recently developed a new model that captures the inherently heterogeneous pore morphology comprising realistic hydrogel matrices, enabling key features of macromolecular transport in the ET regime to be successfully predicted. We can then directly apply these insights using microchip electrophoresis to measure the size dependence of these parameters (particularly diffusion, a seldom characterized parameter).

Here we use this approach to examine electrophoresis of double-stranded DNA in the 100 to 1000 bp size range under conditions favorable for ET-dominated migration. These experiments have led us to identify a range of conditions where a favorable interplay exists between these migration mechanisms, resulting in improved separation performance (e.g., resolving power that increases with DNA size, the opposite of what is conventionally observed). We also explore application of an oscillatory electric field at a specific frequency, in a specific gel morphology where ET effects dominate, to establish a resonance phenomenon that may significantly improve separation resolution. Our microchip electrophoresis experiments indicate that a significant increase in separation resolution occurs over a specific DNA size range relative to the continuous field case. The effect is tunable by adjusting the magnitude of the electric field to vary the window of DNA size where enhancement occurs. These insights may be useful in efforts to develop nanostructured sieving matrices tailored to exploit ET-dominated transport.

See more of this Session: DNA Analysis in Microfluidic and Nanofluidic Devices
See more of this Group/Topical: Topical 3: 2010 Annual Meeting of the American Electrophoresis Society (AES)

(462i) Microfluidic Native Chromatin Immunoprecipitation (μNChIP) Assay for Histone Acetylation Detection

Tao Geng, Department of Agricultural and Biological Engineering, Purdue University, West Lafayette, IN, Ning Bao, Agricultural and Biological Engineering, Purdue University, West Lafayette, IN, Michael Litt, Indiana University School of Medicine - Muncie at Ball State University, Muncie, IN and Chang Lu, Chemical Engineering, Virginia Tech, Blacksburg, VA

Epigenetic modifications, such as acetylation, of the core histone proteins play a role in the control of eukarytotic transcription. To explore the mechanisms, chromatin immunoprecipitation (ChIP) assay is widely applied to studying the presence of various histone modifications at defined genomic regions within living cells or tissues. Native ChIP (NChIP), which uses native chromatin fragments prepared by micrococcal nuclease digestion in the absence of cross-linking, is particularly suitable for detecting histone modifications due to the high affinity between histones and DNA. However, current ChIP assays require a large amount of cells (ideally > 10^7), and involve lengthy and laborious steps (4 days). Here, we developed an integrated microfluidic chip to permit sensitive and rapid NChIP assay based on a small amount of cells. The characteristic of miniaturization allows a low number of cells to be processed on a microfluidic chip while keeping sufficiently high concentrations of chromatin fragments in microchannel. In addition, microscale magnetic beads (ProActive® Protein A Coated Microspheres, 6.2 μ m) with effective reaction surface are used to generate active mixing in order to accelerate the rate of antibody binding onto the target protein, thereby shortening the immunoprecipitation incubation time to 30 min. We assayed histone H3 acetylation at forlate receptor gene of chicken β-globin locus in 6C2 cells using real-time PCR assay. The μ NChIP assay including DNA purification and real-time PCR could be completed in 1 day starting from less than 1000 cells. Our microfluidics-based strategy has the potential to be employed for efficient epigenetic analysis of rare samples such as stem cells, embryonic cells and primary tumors.

See more of this Session: DNA Analysis in Microfluidic and Nanofluidic Devices
See more of this Group/Topical: Topical 3: 2010 Annual Meeting of the American Electrophoresis Society (AES)

(462j) Far-Field Optical Nanoscopiy Based On Continuous Wave Laser Stimulated Emission Depletion (STED) for Nanofluidics

Guiren Wang[1], Cuifang Kuang[2] and Wei Zhao[1], (1)University of South Carolina, Columbia, SC, (2) Zhejiang University, Hangzhou, China

For electrophoresis in a nanochannel, measurement of concentration profile is desirable when Taylor dispersion is concerned. However, due to diffraction limit in optics, it is not possible to measure the concentration profile in a nanochannel with conventional optical method. Here we developed a stimulated emission depletion (STED) microscopy for future concentration profile measurement in nanochannels. STED is one of the breakthrough technologies that are far-field optical microscopy and can break the diffraction limit barrier and achieve nanoscale spatial resolution. We demonstrate a far-field optical nanoscopy based on continuous wave (CW) lasers with different wavelengths, i.e. violet and green lasers for excitation and STED, respectively. Fluorescent dye coumarin 102 is used for validating the depletion efficiency. Fluorescent nanoparticles are selected for characterizing the spatial resolution of the STED system. Linear scanning of the laser beams of the STED system along one line of a glass slide, which is coated with the nanoparticles, indicates that a spatial resolution of about 70 nm has so far been achieved. A two dimensional (2-D) image of the particles pattern of the STED system is constructed and compared with scanning confocal microscope. The present work has further extended the application of the STED microscopy into the blue regime.

See more of this Session: DNA Analysis in Microfluidic and Nanofluidic Devices
See more of this Group/Topical: Topical 3: 2010 Annual Meeting of the American Electrophoresis Society (AES)

(535a) Entrapment of Human Leukemia Cells From Blood Using Contactless Dielectrophoresis

Mike Sano, School of Biomedical Engineering & Sciences, Virginia Polytechnic Institute and State University, Blacksburg, VA, Hadi Shafiee, Department of Engineering Science and Mechanics, Virginia Polytechnic Institute and State University, Blacksburg, VA and Rafael Davalos, School of Biomedical Engineering and Sciences, Virginia Polytechnic Institute and State University, Blacksburg, VA

Dielectrophoresis (DEP), the motion of particles in a non-uniform electric field, has become a robust technique for analyzing cells, micro particles based on their physical and electrical signatures. A new method, contactless Dielectrophoresis (cDEP), shows a good potential to subdue the common trade-offs between sample throughput and selectivity without the need for complicated fabrication processes as evident from conventional sample enrichment techniques. cDEP has been recently used for successful isolation of THP-1 human leukemia monocytes from a heterogeneous mixture of live and dead cells in our laboratories. The absence of contact between electrodes and the sample fluid channel prevent bubble nucleation and avoids any cross contaminating effects the electrodes might have on the sample. In cDEP electrodes are capacitively coupled to the sample containing fluidic channel through dielectric barriers; the application of a high-frequency electric field to these electrodes then induces an electric field in the channel. In order for us to selectively isolate THP-1 tumor initiating cells from blood samples using cDEP with minimal erythrocyte trapping, the Clausius-Mosotti factor of cells was changed using the cell suspension media conductivity as an tuning parameter in the ranges of 100uS/cm2 -300uS/cm2 which in turn is speculated to increase crossover frequency thus aiding in efficient entrapment of the tumor cells from blood. Our preliminary results indicate contactless Dielectrophoresis as an efficient way to detect, isolate and further enrich Circulating Tumor Cells (CTC's) within the biological fluids.

See more of this Session: Electrokinetic Behavior of Microparticles and Nanoparticles: Fundamentals and Applications
See more of this Group/Topical: Topical 3: 2010 Annual Meeting of the American Electrophoresis Society (AES)

(535b) Electrokinetic Vortices and Traveling Waves in Non-Dilute Colloidal Dispersions

Jonathan D. Posner, Chemical Engineering, Mechanical Engineering, Arizona State University, Tempe, AZ and Carlos L. Perez, Mechanical Engineering, Arizona State University, Tempe, AZ

We report a diverse set of coherent fluid flow instabilities, particle patterns, and traveling waves that develop when an ac electric field is applied to nondilute colloidal dispersions with volume fractions that span 3 orders of magnitude. Novel observed phenomena include the following: vortices with steady and unsteady axes of rotation, unsteady time evolution of vortices formation and vortex merging, as well as traveling waves that propagate through the dispersion. Vortical flows are influenced by strong interactions between particle electrical dipoles as well as electric body forces due to electric fields coupled with gradients in particle volume fraction.

See more of this Session: Electrokinetic Behavior of Microparticles and Nanoparticles: Fundamentals and Applications
See more of this Group/Topical: Topical 3: 2010 Annual Meeting of the American Electrophoresis Society (AES)

Application of Capillary Electrophoresis to Predict Crossover Frequency of Polystyrene Particles in Dielectrophoresis

Christian M. White[1], Lisa A. Holland[1], and Parviz Famouri[2]

[1]C. Eugene Bennett Department of Chemistry, West Virginia University, Morgantown, WV 26506

[2]Lane Department of Computer Science and Electrical Engineering, West Virginia University, Morgantown, WV 26506

Dielectrophoresis (DEP) with AC electric fields provides spatial separation of a variety of particles. Depending on the experimental conditions, the particle may respond with positive or negative dielectrophoresis. Under appropriate conditions, this switch from negative to positive dielectrophoresis can be used segregate particles by controlling the frequency. This is easier to accomplish if the permittivity and conductivity of the particles and the aqueous medium suspending the particles are known. The particle conductivity may be determined if zeta potential viscosity, double layer thickness, and the particle radius are known. Capillary electrophoresis instrumentation is useful for measurements of viscosity and electrophoretic mobility of the particles, from which zeta potential is then calculated.[1] Measurements with capillary electrophoresis instrumentation are rapid and sample volumes introduced into the separation capillary range from nano- to femtoliter depending on the separation parameters. The approach is demonstrated with polystyrene particles, which have been subject to different chemical modification. The zeta potential is derived from capillary electrophoresis experiments under different conditions. The DEP behavior of the particles is observed at different frequencies using a DEP cell containing fabricated gold quadrupole electrodes. The particles undergo positive DEP and accumulate near the edges of the quadrupole electrodes used in this study. When the force is negative, the particles undergo negative DEP and collect in the center of the four electrode array. The particle surface conductivities derived from capillary electrophoresis separations differ from those derived from dielectrophoresis experiments by $\leq$ 0.003 Sm^{-1}.[1]

[1]CM White, LA Holland, P Famouri, *Electrophoresis*, **31**(15), pp. 2664-2671, 2010

DIELECTROPHORETIC MANIPULATION OF SILICA NANOPARTICLES IN OPTOFLUIDICS

Aminuddin Kayani[1], Khashayar Khoshmanesh[2], Chen Zhang[1], Arnan Mitchell[1] and Kourosh Kalantar-Zadeh[1]

[1] RMIT University, Melbourne, VIC, Australia.

[2] Deakin University, Geelong, VIC, Australia.

Abstract

We present a dielectrophoretic (DEP) system in microfluidics capable of nanoparticle positioning for the objective of achieving a dense nanoparticle core and primarily liquid cladding optofluidic system. Various DEP conditions were investigated using 450 nm and 230nm diameter silicon dioxide (SiO2) nanoparticles at several alternating current (AC) voltage amplitudes, flow rates and frequencies. Using an array of curved DEP microelectrodes and by utilizing DEP nanoparticle focusing and repelling capabilities, we were able to form a particle dense stream in the center of the microfluidic channel. A high refractive index core composed of DEP assembled nanoparticles with a lower index cladding composed of DI water and very low concentrations of particles was formed. By introducing visible light as the excitation source, it was observed that the nanoparticle aggregate either guides or scatters light. The larger particles tend to scatter light while the smaller particles tend to guide light. The results illustrate a pioneering accomplishment of a configurable nanoparticle-core liquid cladding waveguide that is relevant for optofluidic, lab-on-a-chip and biosensing applications.

Introduction

Nanoparticle positioning techniques have been presented in the past including active and passive manipulation methods in microfluidic systems [1]. In this investigation, we demonstrate that dielectrophoretic (DEP) electrodes are capable of manipulating and controlling nanoparticles in order to form three dimensional assemblies within a microfluidic channel. Coupled with hydrodynamic forces, a network of curved DEP electrodes is used to focus and repel silicon dioxide (SiO2) nanoparticles from the center of a microfluidic channel. This packed stream of nanoparticles forms a region of high refractive index compared to the surrounding media consisting of deionized (DI) water therefore forming a refractive index profile similar to that of a liquid waveguide. Already a vast range of aqueous based optical waveguides utilizing liquid core liquid cladding [2-5] and liquid core solid cladding [6-7] have been demonstrated in the past [8-9]. However, waveguides utilizing nanoparticles as the core and moreover DEP assembled particles have yet to be demonstrated. We present an optofluidic system and a method illustrating the use of DEP assembled nanoparticles for the waveguiding of visible light in microfluidic systems. Using a coherent light source directed along the nanoparticle assembled core, mode profiles at the output were captured. We found that these DEP assembled nanoparticle structures either guide or scatter light at different applied voltages, flow rates and frequencies.

Principles

The process used for the manipulation of nanoparticles in this experiment is known as dielectrophoresis. When a suspended neutral particle is exposed to a non uniform electrical field, electrical charges are induced causing the formation of dipoles within the particles themselves. Both ends of the dipole experience unequal Columbic forces and there is a net translational motion induced on the particle as a result. Depending on its polarizabilities, the particle may experience a force attracting it to the region of high electric field gradients or repelled from it [10-11]. For a spherical particles, the DEP force is given by [10]:

$$\vec{F}_{DEP} = 2\pi r^3 \varepsilon_o \varepsilon_r \,\mathrm{Re}[f_{CM}(\omega)] \nabla E^2_{rms} \qquad [1]$$

where r is the radius of the particle, ε_o=8.854×10^{12} (F/m) is the permittivity of the vacuum, , ε_r is the relative permittivity of the suspending medium, Re[f_{CM}] is the real part of the Clausius–Mossotti (CM) factor, ∇ is the gradient operator and E_{rms} is the root mean square (rms) value of the electric field applied [10]. The real part of the CM factor is given by:

$$f_{CM}(\omega) = \frac{\varepsilon_p * - \varepsilon_m *}{\varepsilon_p * + 2\varepsilon_m *} \qquad [2]$$

where ε_p^* and ε_m^* are the complex permittivities of the particle and the suspending medium respectively given by:

$$\varepsilon * = \varepsilon_o \varepsilon \; - \frac{i\sigma}{\omega} \qquad [3]$$

where $i=\sqrt{-1}$, σ is the conductivity of the medium or the particle and ω is the angular frequency of the applied electric field. The electric field is dependent on the alternating current (AC) voltage applied across the DEP microelectrodes.

The DEP spectrum is denoted by the sign of the Claussius Mossoti factor which is positive when a particle experiences a force attracting it to the region of high electric field gradients or negative when a particle is repelled from the region of high electric field gradients [11]. Simulations of the DEP forces experienced by the nanoparticles independent of size in three axes, DEP-x, DEP-y and DEP-z [12] at various locations throughout the channel are provided (Figure 1). DEP forces are analyzed based on the symmetric electrode pattern on one side, at specific microelectrode locations and channel heights (Figure 1(a)).

Ignoring the effect of flow rates on particle manipulation, the particles are subject to the highest electric field gradients when they flow in levitation at heights between 15 to 40 μm. For z=10μm, the particles experience the maximum DEP force along the y-axis (Figure 1(b) and (d)). Particles on the left of the microelectrodes (y = 0 to 25 μm) are repelled (negative DEP) to the side walls while particles above the electrode region (y = 25 to 60 μm) are pushed towards the centerline (positive DEP) by DEP-y forces (Figure 2(b)-(e)). Particles experience positive DEP-z force throughout the channel, i.e. they experience particle trapping throughout the region of high electric field gradients (Figures 2(b)-(e)). Choosing the appropriate AC potential and a microfluidic flow rate that allows laminar flow is fundamental to enable particle focusing along the center band of the channel. Low magnitude signals do not efficiently trap particles in the center while overly polarized particles are levitated and repelled to the sidewalls rather than be focused.

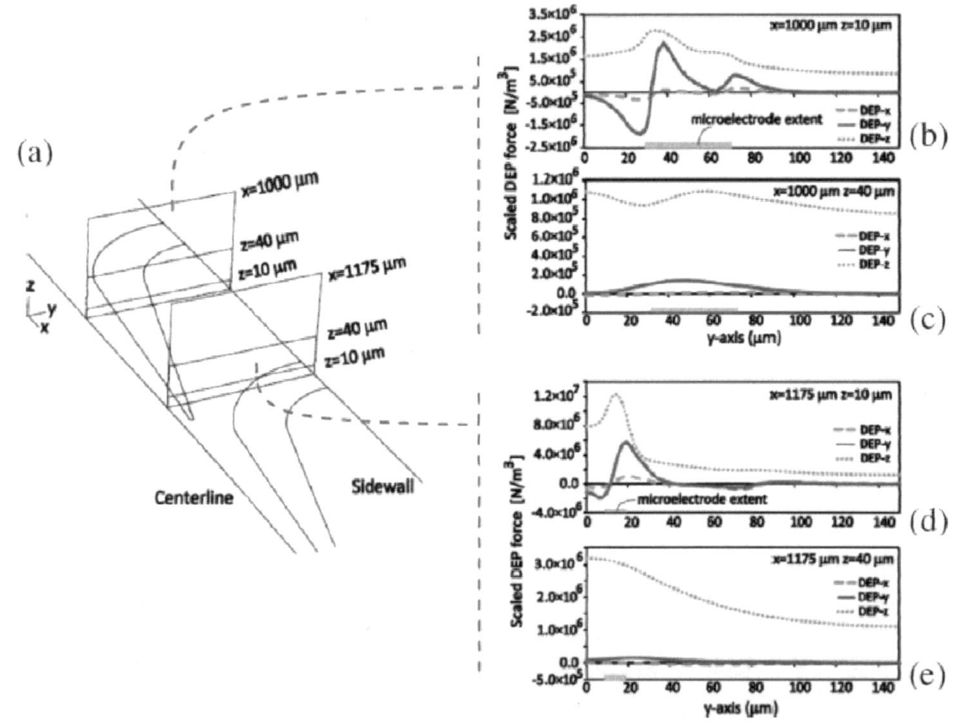

Figure 1: (a) Electrode analysis locations, DEP force for x=1000 μm and (b) z=10 μm (c) z=40 μm, x=1175 μm and (d) z=10 μm (e) z=40 μm

Fabrication and Experiment

The DEP microelectrodes were first designed using Advanced Design System 2009 and a chrome mask fabricated to enable photolithography of the curved electrode patterns. Figure 2 (a-c) shows the array, structure and dimensions of the curved DEP microelectrodes fabricated on the mask. Microtip and finger pair microelectrode pairs [11,13] did not provide a smooth funneling effect of nanoparticles and therefore were not selected for this investigation. The system was fabricated on a glass microscope slide. The slide was deposited with Cr:Au (50nm:150nm) by electron beam deposition. The microelectrodes were patterned using photolithographic techniques [14]. Initially, AZ1512 (Microchemicals) was spun using a wafer spinner (Karl Suss RC8) at 500rpm, 100rpm/sec acceleration, for 5 seconds for a spread coat followed by 3000rpm, 300rpm/sec acceleration for 30 seconds. The slide was oven baked for 20 minutes at 90 deg C followed by an exposure dose of 125mJ/cm^2 (MJB3 mask aligner). The slide was then developed using AZ 1:4 photoresist developer followed by etching using Aqua Regia and Chromium etchants for gold and chromium respectively.

The microfluidic channel was 500 μm x 1.7cm x 100 μm fabricated using polydimethylsiloxane (PDMS) soft lithography[15]. A cast master was made using SU8 2050 (Microchem) spun at 500rpm, 100rpm/sec acceleration, for 5 seconds for a spread coat followed by 2000rpm, 300rpm/sec acceleration for 30 seconds to obtain a 70 μm feature height. The master was then soft baked on hotplate at 65 deg C for 3 minutes followed by 95 degrees Celcius for 9 minutes. Next the wafer master was exposed with a dose of 200 mJ/cm^2 using a film mask and post exposure baked for 2 minutes at 65 degrees C and 7 minutes at 95 degrees C. The A 1:10 mixture of PDMS curing agent: elastomer (Sylgard 184) was mixed, degassed and poured into microfluidic shims and cured at 75 deg C for 2 hours in an oven. Prior to attachment to the microscope slide, the microfluidic reservoirs were punched with 0.75 mm holes and plasma treated for 2 minutes to make the surface more hydrophilic prior to attachment on the slide. Post attachment, microfluidic tubing were inserted in both reservoirs and connected to a Luer interfaced 5ml syringe mounted on a syringe pump (Harvard Apparatus PHD 2000).

A series of experiments were conducted using 230 and 450 nm silica particles by changing the applied voltage and frequency in order to study the effect of these parameters in establishing and tuning the nanoparticle core microfluidic waveguide. The SiO2 particles (Microspheres Nanospheres) were imaged using a scanning electron microscope (SEM) to assure uniformity of particle size (Figure 2(d) and Figure 2(e)). The flow rate chosen was 5 μL/min flow rate which was found suitable to allow formations of a dense nanoparticle stream with minimal turbulent mixing [16] and without jeopardizing the ability of DEP forces to position the nanoparticles effectively. A 70 μm groove was included in the PDMS cast master in order to insert a cladding stripped multimode fiber core of 62.5 μm diameter. This fiber core was pre-aligned to the center of the DEP microelectrode spacing and a transparent window of the PDMS was made for profiling the laser output. It was polished by curing it against the surface of a silicon wafer. Subsequently, a visible light source at λ = 635 nm (OZ Optics) was used as the light source and a profiler (Philips SPC900NC) was used to capture the output (Figure 2(f)). The system was subsequently flushed with 1-propanol (IPA) before being infiltrated with a 4 % volumetric concentration of 230nm and 450nm suspended silica particles in DI water.

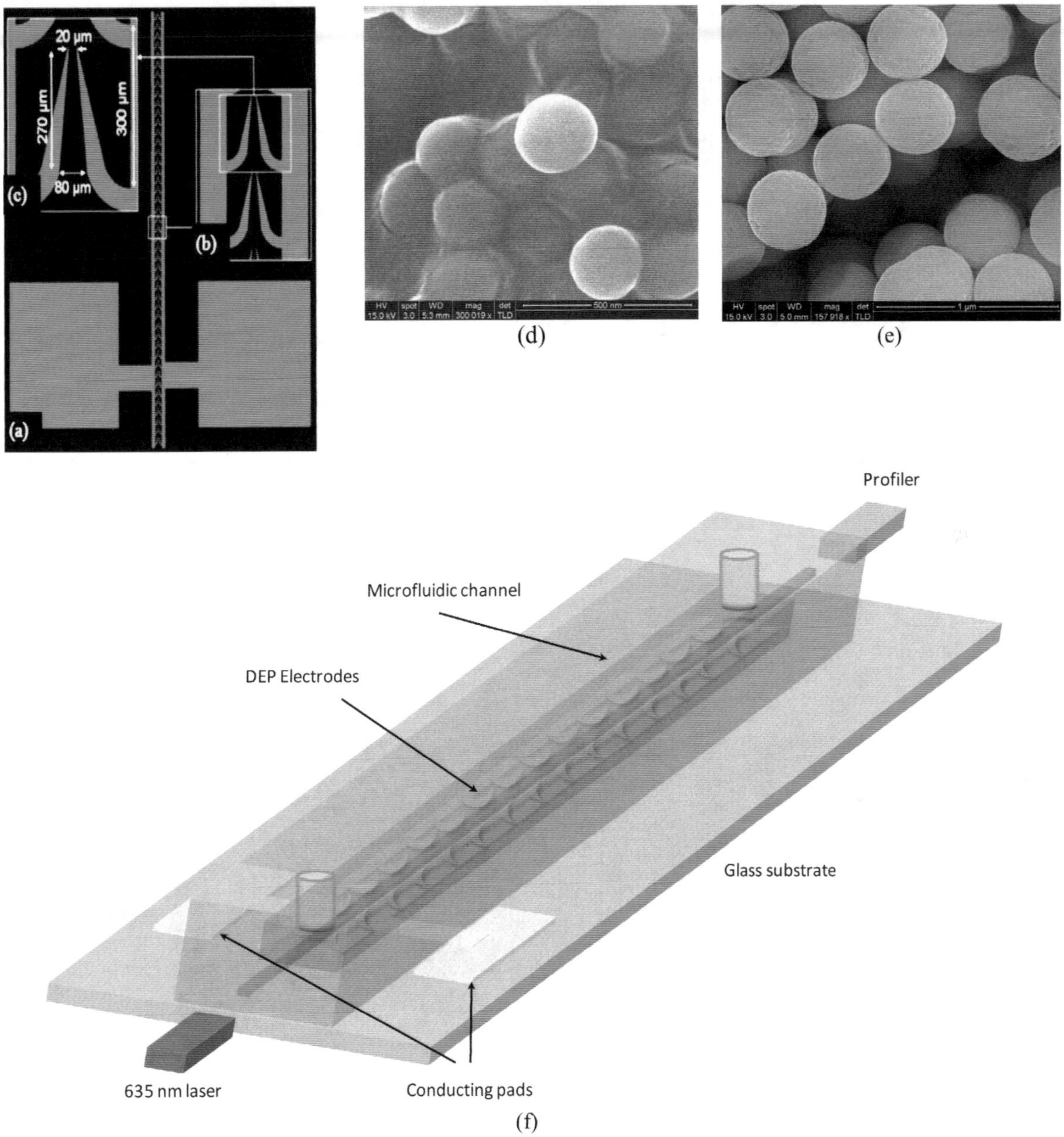

Figure 2: (a) DEP microelectrode array (b) Curved oblique electrode design (c) Structure dimensions (d) SEM image of 230nm SiO2 particles (e) SEM image of 450 nm SiO2 particles (f) Optofluidic device integrated with DEP used for profile measurement

Results Analysis

Using the inverted microscope, several significant DEP occurrences were captured (Figure 3). For 450 nm SiO2 particles, we noticed that at a frequency of 300 kHz, the particles were trapped in the center (Figure 3(a)). This particle trapping condition was prevalent as we increased the frequency until 800 kHz when a particle stream was formed. As the frequency reached 1.5 MHz, the nanoparticle dense stream was formed (Figure 3(b)). Depending on the magnitude of the AC potential applied, the nanoparticles were either trapped, moving slowly or flowing in laminar with the aid of external hydrodynamic forces. For frequencies above 1.5 MHz, the particles began to form a funnelled stream of particles between the electrodes forming a region dense with particles in the centerband (Figure 3(b)). The narrowband appeared brighter than the surrounding regions due to scattering of impinged light between the particles. For the 230 nm particles, a similar pattern was observed as we increased frequency from 250 kHz to 1.5 MHz in that particles formed a smoother stream at higher frequencies. However, the centerband appeared darker at 1.5 MHz.

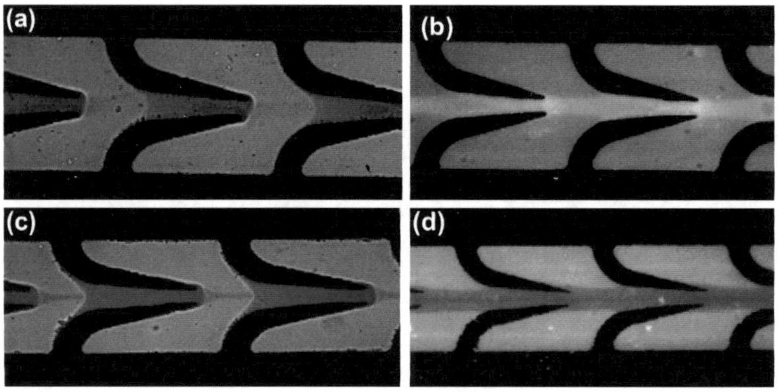

Figure 3: DEP assembly and repulsion of SiO2 nanoparticles at flow rates of 5 μL /min and AC potential of 15 Vpp. Frequencies of (a) 300 kHz, (b) 1.5 MHz for 450 nm, (c) 300 kHz, and (d) 1.5 MHz for 230 nm particles.

The output profiles were then investigated (Figure 4) using various AC applied voltages. We found that by increasing the voltage amplitude, the width of the nanoparticle stream decreased indicating efficient particle trapping between the electrodes. At 635 nm wavelength, as the refractive index of water is n_{Df}=1.33 at room temperature while the refractive index of silica is n_{si} = 1.47[17]. The index of silica particles focused by DEP therefore lies in the range of 1.33 < n_{si} < 1.47 subject to the degree of concentrations in the centerband. Therefore, DEP focused conditions hypothetically establishes a profile across the channel that satisfies total internal reflection conditions for waveguiding. The refractive index contrast therefore varies between $0 < \delta n < 0.14$ and changes as a function of DEP conditions, with a distinct refractive index profile when a dense nanoparticle stream is formed where, $\delta n = n_{si} - n_{DI}$. Our observation was the 230 nm particles focused in the centerband seem to allow the passage of light through the microfluidic channel. Conversely, the 450 nm particles while being focused at the centerband, seem to completely block out the passage of light through the microfluidic channel by scattering.

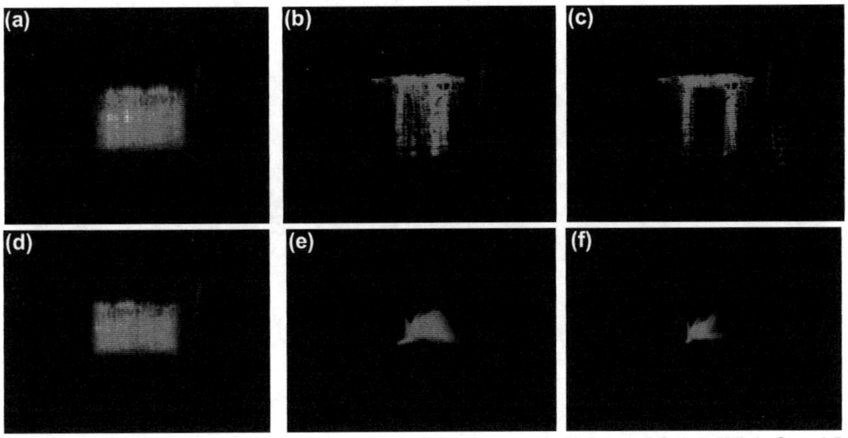

Figure 4: Output profile observed under DEP trapping conditions (f=1 MHz) for 450 nm silica nanoparticles at (a) 0 V, (b) 7.5 V, (c) 15 V, and 230 nm particles at (d) 0 V (e) 7.5 V and (f) 15 V.

We correlate the results with the Lorenz Mie scattering theory[18] and we attribute scattering and waveguiding due to the spacing between the particles themselves when they are trapped in close contacts with each other. The spacing between larger particles i.e. 450 nm silica is considerably larger than the inert particle spacing of 230 nm particles. As a result, 450 nm particles tend to scatter light and block out the passage of light completely (Figure 4 (d) – (f)) as compared to the 230 nm particles which are capable of guiding light (Figure 4 (a) – (c)). Numerically, we can estimate the inter particle spacing in terms of volume of tightly trapped 450 nm and 230 nm to be a volume of a 230 nm and 120 nm spheres [12] respectively. Considering that light scattering decreases for particles with diameter, $d_{particle} < \lambda/4$, we can infer that 230 nm are able to guide light more effectively due to its ability to be tightly packed to each other under DEP conditions (Figure 5 (a)). Moreover, the scattering efficiencies based on a 635 nm wavelength source are 0.043 and 0.185

(http://omlc.ogi.edu/calc/mie_calc.html) for 230 nm and 450 nm particles respectively. This also indicates that 450 nm particles has 4.3 times more capacity [17] to scatter light than 230 nm particles (Figure 5 (b)).

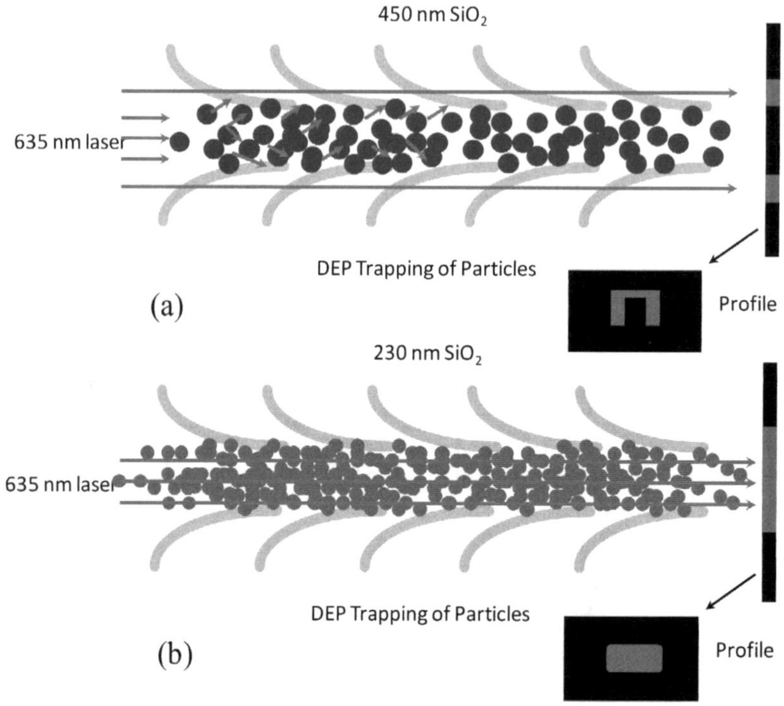

Figure 5: Waveguiding and scattering as a result of DEP: (a) 230-nm particles (b) 450-nm particles.

Conclusion

The application of suspended nanoparticle waveguides based on DEP forces is possible using a well defined electrical force, frequency and by externally manipulating hydrodynamics of the microfluidic channel, one is capable of forming 3D tuneable optical elements within a channel. This paper demonstrates the implementation of integrating an optofluidic system with DEP for the purpose of producing tuneable waveguiding structures in liquid. Scattering and waveguiding was demonstrated possible and it was varied as a function of DEP settings. Further research work in this area could include attempting different nanoparticles and changing the DEP electrode geometries. For example the use of semiconducting nanoparticles may provide interesting results and a travelling wave DEP system may provide more precision particle sorting and manipulation capabilities. This work further illustrates the potential of integrating nanomaterials in microfluidics for the purpose of biosensing, flow cytometry, optical waveguiding and lab on chip applications.

References

1 Xuan, X., Zhu, J. & Church, C. Particle focusing in microfluidic devices. *Microfluidics and Nanofluidics* **9**, 1-16 (2010).

2 Bernini, R., De Nuccio, E., Minardo, A., Zeni, L. & Sarro, P. Liquid-core/liquid-cladding integrated silicon ARROW waveguides. *Optics Communications* **281**, 2062-2066 (2008).

3 Dress, P. & Franke, H. A cylindrical liquid-core waveguide. *Applied Physics B-Lasers and Optics* **63**, 12-19 (1996).

4 Li, X. C. *et al.* A liquid waveguide based evanescent wave sensor integrated onto a microfluidic chip. *Applied Physics Letters* **93**, 193901-193901 to 193901-193902 (2008).

5 Wolfe, D. B. *et al.* Dynamic control of liquid-core/liquid-cladding optical waveguides. *PNAS* **101**, 12434-12438 (2004).

6 Risk, W. P., Kim, H. C. & Miller, R. D. Optical waveguides with an aqueous core and a low index nanoporous cladding. *Optics Express* **12**, 6446-6455 (2004).

7 Schueller, O. J. A., Zhao, X. M., Whitesides, G. M., Smith, S. P. & Prentiss, M. Fabrication of liquid-core waveguides by soft lithography. *Advanced Materials* **11**, 37-41 (1999).

8 Hawkins, A. R. & Schmidt, H. Optofluidic waveguides: II. Fabrication and structures. *Microfluid Nanofluid* **4**, 17-32 (2008).

9 Wolfe, D. B., Vezenov, Dmitri V., Mayers, Brian T., Whitesides, George M., Conroy, Richard S. and Prentiss, Mara G. Diffusion-controlled optical elements for optofluidics. *Applied Physics Letters* **87**, 181105-181101 to 181105-181103 (2005).

10 Pohl, H. *Dielectrophoresis, The behavior of neutral matter in nonuniform electric fields*. 1st edn, (Cambridge University Press, 1978).

11 Zhang, C., Khoshmanesh, K., Mitchell, A. & Kalantar-zadeh, K. Dielectrophoresis for manipulation of micro/nano particles in microfluidic systems. *Analytical and Bioanalytical Chemistry*, 1-20 (2009).

12 Kalantar-Zadeh, K., Khoshmanesh, K., Kayani, A. A., Nahavandi, S. & Mitchell, A. Dielectrophoretically tuneable optical waveguides using nanoparticles in microfluidics. *Applied Physics Letters* **96** (2010).

13 Zhang, C. *et al.* 85-94.

14 Zheng, L., Shengdong, L., Brody, J. P. & Burke, P. J. Manipulating Nanoparticles in Solution with Electrically Contacted Nanotubes Using Dielectrophoresis. *Langmuir* **20**, 8612-8619 (2004).

15 McDonald, J. C. *et al.* Fabrication of microfluidic systems in poly(dimethylsiloxane). *Electrophoresis* **21**, 27-40 (2000).

16 Khoshmanesh, K. *et al.* Dielectrophoretically assembled particles: feasibility for optofluidic systems. *Microfluidics and Nanofluidics* **9**, 755-763 (2010).

17 Kayani, A. A. *et al.* Novel tuneable optical elements based on nanoparticle suspensions in microfluidics. *Electrophoresis* **31**, 1071-1079 (2010).

18 Mishchenko, M., Travis, L. & Lacis, A. *Scattering, absorption and emission of light by small particles* 2edn, (Cambridge University Press, 2002).

DIELECTROPHORETIC MANIPULATION OF MICROORGANISMS IN AN ARRAY OF INSULATORS: THEORETICAL AND EXPERIMENTAL RESULTS

Héctor Moncada-Hernández[1], Javier Baylon-Cardiel[1], Blanca H. Lapizco-Encinas[2,*]
Email: blapizco@cinvestav.mx

[1]*BioMEMS Research Chair, Tecnologico de Monterrey, Monterrey, Mexico*
[2]*Microscale Bioseparations Laboratory, CINVESTAV-Monterrey, Apodaca, Mexico*

Abstract

Dielectrophoresis (DEP) has been proved as an efficient separation technique with a great potential for miniaturization. DEP is defined as the motion of dielectric particles due to a force exerted on them originated by the presence of a nonuniform electric field. Electrode-based DEP allows obtaining high electric field gradients with low applied voltages presenting several disadvantages. An alternative of this electrodes approach is the implementation of an array of insulating structures to achieve the nonuniform electric fields with two electrodes that straddle the array of structures. This approach is known as insulator-based dielectrophoresis (iDEP). This work presents the manipulation and separation of viable and non-viable cells achieved, by employing a microdevice for iDEP. A computational model is employed to predict the effect of DEP on this bioparticles; the model consists on the implementation of Finite element analysis on a geometry which consists of an array of insulating circular posts distributed along a microchannel. Successful manipulation and separation of viable and non-viable *S. cerevisiae* was evaluated experimentally employing DC electric fields inside a microdevice containing an array of insulating structures. The results from simulation and experiment were compared.

Introduction

The significant advance in microfabrication technology in the past two decades has allowed the development of the so called Micro-Electro-Mechanical Systems, or MEMS; and these systems in turn have set the basis for the miniaturization of analytical devices with biological applications (BioMEMS). Due to their sizes and the advantages they presents, BioMEMS are being studied for diverse applications, such as medical diagnose and biological assays. This interest is due to the advantages that come with working in the microscale: shorter separation times, lower production costs, higher performance and smaller required quantities of samples and reagents.

For this purpose, the implementation of Dielectrophoresis (DEP) has been proved to be an efficient technique with a great potential for miniaturization. DEP is the motion of dielectric particles due to a force exerted on them, originated by the presence of a nonuniform electric field.[1] Traditionally, DEP has been carried out employing arrays of microelectrodes and alternate current (AC) electric fields. Electrode-based DEP allows obtaining high electric field gradients employing low applied voltages.

Cummings and Singh proposed a different approach for the implementation of DEP.[2] They employed an array of cylindrical posts embedded in a microchannel made with an electrical insulating material. This approach is known as insulator-based DEP (iDEP). In iDEP, the electric potential is applied across an array of insulating structures. The exerted DEP force on a particle depends greatly on its dielectric properties, as well as on its shape. Due to their complexity, biological particles can be regarded as particles consisting of several layers of different size, each one having different dielectric properties.[3] Moreover, the shape of the particle can be approximated employing different geometries, such as spheres and ellipsoids.[4]

Important mathematical modeling efforts have been devoted to further develop insulator-based dielectrophoretic microdevices. Kwon *et al.* investigated the electrokinetic (EK) and DEP forces exerted on particles in a microchannel employing numerical simulations and proposing an improved geometry to increase particle transport in EK streamlines.[5] Chen *et al.* studied the mechanism of particle transport and separation for a sawtooth channel. Since motion of single particle in a DC-iDEP device was not fully understood, their study was limited to a single infinitesimal particle.[6] Baylon-Cardiel *et al.* studied the manipulation of particles employing DC and AC signals. Initially, applying a DC signal to immobilize and concentrate particles in bands in order to be further manipulated with an AC signal with low frequencies.[7]

Several studies on the implementation of iDEP for the manipulation and concentration of biological particles have been made. Chou and Zenhausern successfully achieved concentration, separation and lyses of cells in a microdevice with a pair of triangle hurdles which make a constriction of the channel, employing AC electric fields.[8] Zhou and Suehiro *et al.* developed a filter employing glass beads as insulating structures to create a nonuniform electric field.[9, 10] Cho *et al.* employed a membrane as an insulating structure and *E. coli* cells are forced to flow through it; the cells are then attracted to the regions of high field intensity due to the effect of DEP.[11] Lapizco-Encinas *et al.* successfully separated and concentrate *E. coli*, *B. subtilis*, *B. cereus*, *B. megaterium* and Tobacco Mosaic Virus in an array of insulating posts employing DC electric fields.[12-14] Pysher and Hayes simultaneously concentrated and separated viable and non-viable *B. subtilis*, *E. coli* and *S. epidermis* cells.

In the present work, the manipulation and separation of different biological particles is achieved, by employing a microdevice for iDEP. A computational model is employed to predict the effect of DEP on the particles; the model consists on the implementation of Finite element analysis on a geometry which consists of an array of insulating circular posts distributed along a microchannel. Furthermore, the manipulation and separation of viable and non-viable *S. cerevisiae* cells was evaluated experimentally employing DC electric fields inside a microdevice containing an array of insulating structures. The results from simulation and experiment were compared

Materials and Methods.

Saccharomyces cerevisiae ATCC 24858 were cultured in YM Broth (YMB) at 30° C in a shaker incubator for 18 h to achieve late log phase to a cell concentration of $6x10^6$ cells/ml verified by OD measurements at 600 nm. Cells were centrifuged and re-suspend in DI water to wash away the culture media. Cells were labeled with Syto® 9 bacterial stain (Invitrogen, Carlsbad, CA) that fluoresce green (excitation/emission 483/503 nm). Labeled cells were washed with DI water to remove any excess dye and finally re-suspended in the buffer solution to reach the desired cell concentration ($6x10^8$ cells/ml).

Cells were exposed to high temperatures in order to get non-viable cells to evaluate dead cells performance. Dead cells were obtained placing live cells in a small glass bottle with screw cap inside a water bath at a constant temperature of 80° C for 20 minutes. Dead cells were then labeled with Propidiun Iodide (Invitrogen, Carlsbad, CA), which will fluoresce red (excitation/emission 490/635). Dead labeled cells were then washed following the same protocol as live cells.

A 2% weight of Albumin, bovine serum (Sigma-Aldrich, St. Louis, MO, USA) solution in DI water was prepared as suspending media, resulting in a pH of 6.3 and a conductivity of 150 μS/cm. The conductivity and pH were measured with a multiparameter bench meter HI 255 (Hanna Instruments, Woonsocket, RI, USA) and a Orion 3 Star Benchtop pH meter (Thermo Scientific, Beverly, MA, USA), respectively.

A schematic representation of the experimental setup is shown in Figure 1. Experiments were conducted in a microfluidic chip consisting of two Schott D263 glass wafers (50.8 x 50.8 mm length, 1.1 mm thick, Howard Glass, Worcester MA, USA) using standard photolithography, wet etch and bonding techniques. Holes were drilled in the second wafer and then bonded with the previous wafer in order to get inlet and outlet reservoirs for the microchannels.

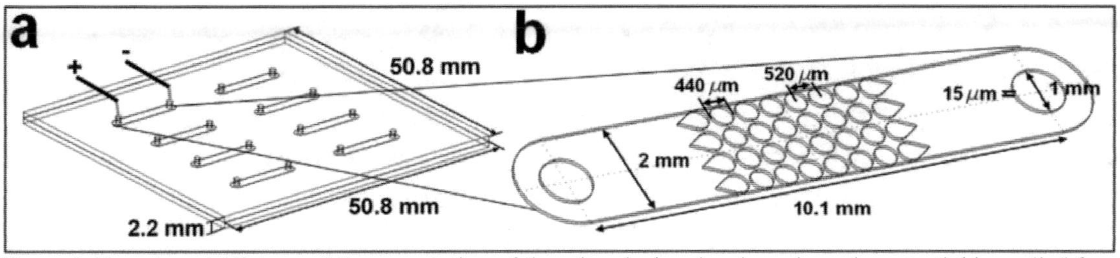

Figure 1. Experimental Set-up. a) Representation of the microdevice showing where the potential is applied for a selected microchannel. b) Representation of microchannel with the insulating posts array

The microdevice contained eight microchannels, each microchannel was straddled by an inlet and outlet reservoirs. The microchannels were 15-µm-deep, 10.1-mm-long and 2-mm-wide and contained an array of cylindrical posts from the same glass material (Figure 1). The posts had a diameter of 440 µm and were arranged 520 µm center-to-center distance of. The post tranverse the entire depth of the microchannel (Figure 1b). The post array consisted of 32 posts arranged in 8 rows of 4 posts each.

A high voltage sequencer model HVS448 (LabSmith, Livermore, CA USA) was used to apply DC electric fields. Visualization of dielectrophoretic behavior was performed employing an inverted epifluorescence video microscope for microfluidics model SVM340 (LabSmith, Livermore, CA USA). The microscope is equipped with a light module with a configuration of two colors LEDs, blue light that excites at a wavelength of 480 nm and red light exciting at 625 nm; which is appropriated for detecting the Syto® 9, Syto® 17 and Propidium Iodide fluorescent stains employed in this work. For the operation of the high voltage sequencer and the microfluidics microscope is required the use of a personal computer.

Mathematical model and theoretical background

The geometry of the iDEP device described in the experimental setup section was captured with *COMSOL Multiphysics* software. A triangular mesh was built in order to solve the Laplace equation using the *Lagrange element* implemented in the software. Finite element analysis was applied in a triangular mesh consisting of 600 elements and 1800 nodal points. The electric potential V was approximated by a polynomial of low order at each mesh point. From the numerical solution it is possible to obtain values for the parameters describing the time dependant motion of the particles, such as the electrical field $\bar{E}$ and the gradient of E^2.

Electrokinetics

The motion induced on a particle due to the electrical force exerted on it can be expressed in terms of the time dependant electrokinetic velocity (*EK*) as:

$$\vec{u}_{EK} = \mu_{EK}\vec{E} \qquad (1)$$

Where μ_{EK} is the electrokinetic mobility and $\vec{E}$ is the spatially non uniform electric field.

The mobility μ_{EK} takes into account the contribution of electrophoretic (EP) motion of the particle and the electroosmotic (EO) motion of the suspending medium in terms of their respective mobilities as:

$$\mu_{EK} = \mu_{EP} + \mu_{EO} \qquad (2)$$

Dielectrophoretic force

The dielectrophoretic force on a spherical particle of radius R exerted by an electric field with magnitude E is defined as

$$\vec{F}_{DEP} = 2\pi R^3 \varepsilon_m f_{CM} \nabla E^2 \qquad (4)$$

Where R is the particle radius, ε_m is the permittivity of the suspending medium and f_{CM} is the Classius-Mossotti (CM) factor, which accounts for particle polarizability. The CM factor for an applied DC potential[22] is defined as:

$$f_{CM} = \left[\frac{\varepsilon_p^* - \varepsilon_m^*}{\varepsilon_p^* + 2\varepsilon_m^*}\right] \qquad (5)$$

where σ_p and σ_m are the conductivities of the particle and the suspending medium, respectively. A biological particle, such as a cell, is a complicated structure, as it contains several different layers, each with different dielectric properties. A simple model for such particle consists of a spherical particle (the cell interior, with conductivity σ_{int} and radius R_{int}), surrounded by a concentric dielectric shell (the cell membrane, with conductivity σ_{mem} and radius R_{mem}). In the DC case, the effective conductivity of the particle, σ_p, can be expressed as

$$\sigma_p = \sigma_{mem}\left[\frac{\left(\frac{R_{mem}}{R_{\text{int}}}\right)^3 + 2\left(\frac{\sigma_{\text{int}} - \sigma_{mem}}{\sigma_{\text{int}} + 2\sigma_{mem}}\right)}{\left(\frac{R_{mem}}{R_{\text{int}}}\right)^3 - \left(\frac{\sigma_{\text{int}} - \sigma_{mem}}{\sigma_{\text{int}} + 2\sigma_{mem}}\right)}\right] \qquad (6)$$

Moreover, the contribution of the DEP force to the motion of the particle can be expressed in terms of the dielectrophoretic velocity:

$$\vec{u}_{DEP} = -\mu_{DEP} \nabla E^2 \qquad (7)$$

where μ_{DEP} is the dielectrophoretic mobility. For a spherical particle of radius R, the mobility μ_{DEP} is defined as:

$$\mu_{dep} = \frac{R^2 \varepsilon_m f_{CM}}{6\eta} \qquad (8)$$

where η is the suspending medium viscosity.

Dielectrophoretic trapping

In order to achieve immobilization of a particle due to DEP, its contribution to the particle motion must overcome other transport mechanisms, such as diffusion and EK motion. In this study, it is assumed that the main contributions to the flux of particles $\vec{j}$ along the microchannel come from EK and DEP, as:

$$\vec{j} \approx \vec{u}_{EK} + \vec{u}_{DEP} \qquad (9)$$

It can be assumed that particle immobilization is achieved in the regions were the condition $\vec{j} \cdot \vec{E} = 0$ is satisfied. In terms of the EK and DEP mobilities, this condition can be expressed as:

$$(\mu_{EK}\vec{E}-\mu_{DEP}\nabla E^2)\cdot\vec{E}=0 \qquad (10)$$

Moreover, in terms of a time dependant electric field, the condition for trapping can be expressed as function of time as:

$$\frac{\mu_{DEP}\nabla E^2}{\mu_{EK}E^2}\cdot\vec{E}>1 \qquad (11)$$

Results and discussion

There are three main mechanisms affecting the kinetic behavior of particles in a microfluidic device under a DC electric field: electrophoresis, electroosmotic flow and dielectrophoresis. Due to the size of the particles presented, electrophoresis can be neglected. Electrokinetic mobility depends only of the electroosmotic mobility, determined experimentally for the suspending medium employed. The information of dielectric properties that was fed to the model is presented in table 1.

Table 1. Dielectric properties of viable and non-viable *S. cerevisiae* cells.

Property	Viable cells	Non-viable cells
Cytoplasm conductivity	0.5 S/m	0.05 S/m
Cytoplasm permitivity	50	50
Membrane conductivity	$2.5\text{x}10^{-7}$ S/m	$1.6\text{x}10^{-4}$ S/m
Membrane conductivity	10	10

A sample of viable *Saccharomyces cerevisiae* cells was placed in the inlet reservoir and a DC potential was applied in order to observe the dielectrophoretic response of cells in the insulating structures array. Figure 2a and b show yeast cells under and electric field of 600V and the simulated trapping zones under the same field, respectively. Viable cells are exhibiting negative DEP and are being repelled from the zones of higher electric field intensities, thus being concentrated in bands just before the constriction created by the insulating posts. Trapping zones showed in Figure 2b shows that a trapping condition between 3.5 and 4 is needed to immobilize viable yeast cells with the present geometry and electric field. The image of condition for trapping (Fig. 2b) shows the zone at which cells are completely trapped, the closest to the constriction between the posts. Experimentally, yeast cells are repelled from the narrower zones almost meeting the trapping condition.

Analogous with the viable yeast cells, non-viable yeast cells were placed in the microdevice under an electric field of 1100 V/cm. In Figure 3a can be observe dielectrophoretic trapping by immobilization of the dead cells by negative DEP. The simulation presented in Figure 3b shows trapping zones where the trapping condition for dead yeast cells trapping is between 2 and 4 to achieve immobilization of cells and therefore a cells concentration in every row of the array of insulating structures.

It can be observed from Figures 2 and 3 that dielectrophoretic trapping and concentration can be achieved at different electric fields for viable and non-viable cells. Since membrane conductivity decreases considerably when a cell is non-viable, dielectrophoretic trapping can be selective between viable and non-viable cells. It can be seen that trapping zones differ from one another since viable cells experience a more strong negative DEP force, being pushed farther away from the constriction of the channel.

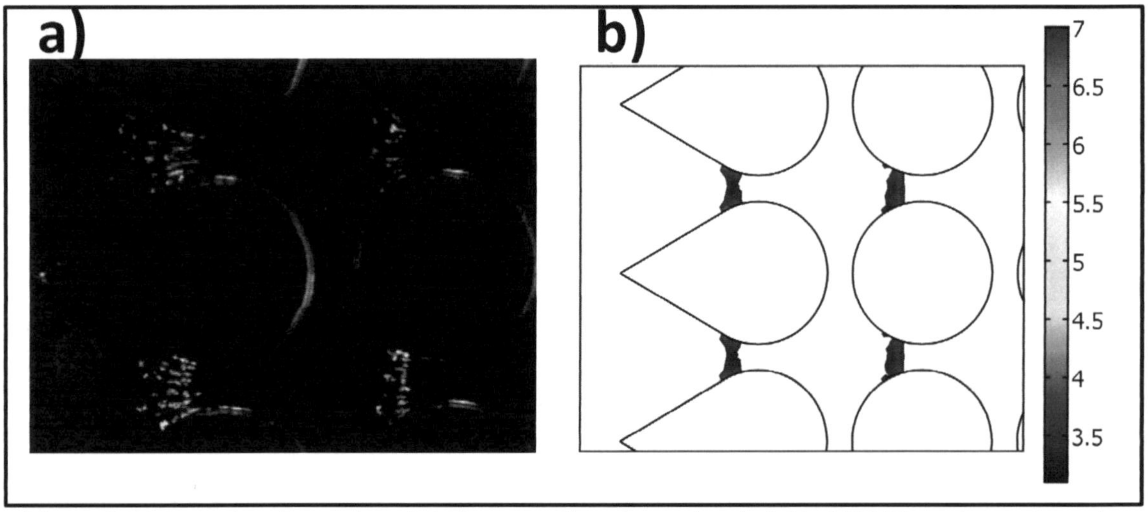

Figure 2. Dielectrophoretic trapping of viable yeast cells under an electric field of 600 V/cm. a) experimental trapping of dyed yeast cells. b) Simulation of trapping zones employing the equation of condition for trapping.

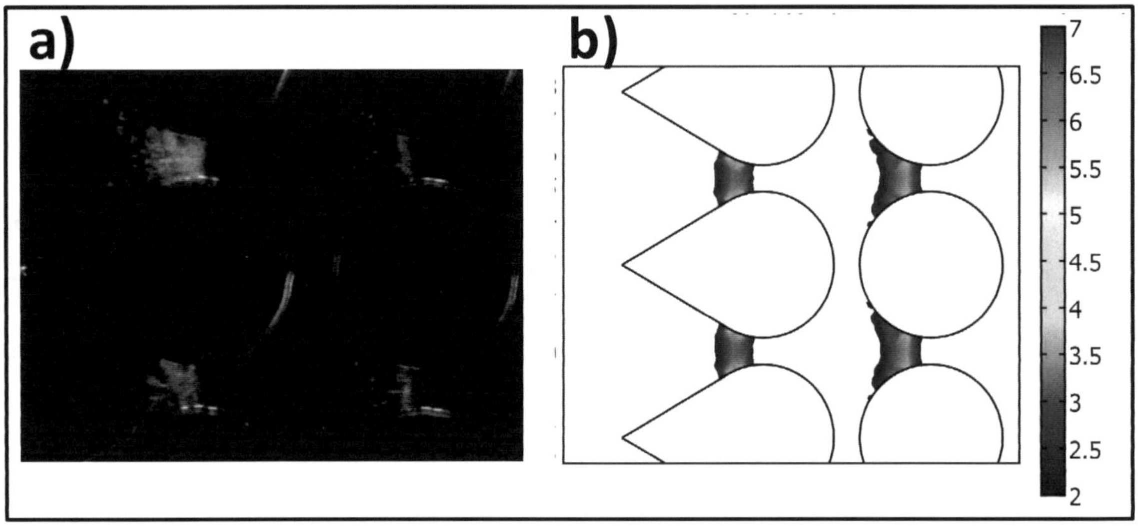

Figure 3. Dielectrophoretic trapping of non-viable yeast cells under an electric field of 600 V/cm. a) experimental trapping of dyed yeast cells. b) Simulation of trapping zones employing the equation of condition for trapping

Based on the results obtained from Figure 2 and 3, a separation of viable and non-viable yeast cells was carried out experimentally and simulated with the model developed. Figure 4 shows the trapping of a mixture of live and dead cells in two bands of dielectrophoretic trapping and the simulated trapping zones for both cells. In Figure 4a, negative dielectrophoretic trapping of a mixture of viable and non-viable *S. cerevisiae* cells is shown. It can be observed that two bands forms in the second row of insulating structures, corresponding the one of the left to viable yeast cells (in red circle in Fig. 4a) and the right to non-viable yeast cells. This can be explained as follows, since viable yeast cells experience a greater DEP force than non-viable, thus being pushed farther away from the zones of higher electric field. Bands in the first row don't seem to be well defined, it is due to cells concentration and that all cells tend to concentrate and push cells forward due to electrokinetic flow and being pushed back due to dielectrophoretic force. In contrast with this, in the second row, effects of cells pushing forward is not as notorious since cells concentration is not as considerable. From Figure 4b it can be observed the two zones of trapping for an electric field of 750V/cm. Blue zones for viable yeast cells and red for non-viable cells are shown near the constrictions of the channel. It is important to notice that trapping zone for non-viable cells is closer to the zones of higher electric field intensity, while the zone for viable cells are farther away from the constriction. This is consistent with what it can be observed from Figure 4a.

It is also important to consider that the difference between experimental trapping bands, and simulated trapping zones in Figure 4 are due mainly because of the concentration of cells. The model does not consider the presence of cells in the system that may influence in the electric field. Physically, concentrated cells occupy a volume and have a different conductivity of that of the suspending medium. The presence of non-viable cells in a concentrated band contributes in the effect of having the viable cells pushed even farther than when they are on their own.

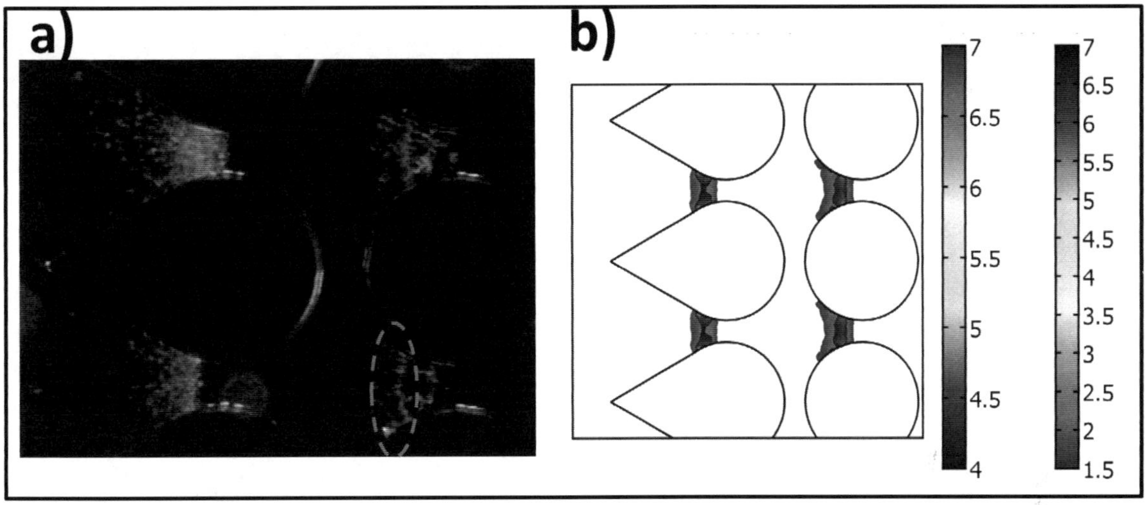

Figure 4. Dielectrophoretic trapping of a mixture of viable an non-viable yeast cells under an electric field of 750 V/cm. a) experimental trapping of dyed yeast cells. b) Simulation of trapping zones for viable (blue) and non-viable (red) yeast cells employing the equation of condition for trapping

Conclusions

In this work the great potential for dielectrophoresis as a method for the manipulation of particles was demonstrated. The results obtained here consisted in the simulation of the dielectrophoretic response of viable and non-viable *S. cerevisiae* cells independently and the separation of a mixture of these live and dead cells. The use of mathematical modeling, as a tool to predict electrokinetic behavior of particles in microchannels, provide information about the design of new and optimized microdevices for specific needs. Differences between experimental and mathematical simulation are due to the presence of particles in the system that are not considered in the current model. There is still work to be done in the field of simulation of electrokinetic particle behavior; where parameters such as cells concentration in the zones of dielectrophoretic trapping should be included.

References

1. M. M. Meighan, S. J. R. Staton and M. A. Hayes, *Electrophoresis*, 2009, **30**, 852-865.
2. E. B. Cummings and A. K. Singh, in *Proceedings of SPIE: Conference on Microfluidic Devices and Systems III*, ds. C. H. Mastrangelo and H. Becker, Santa Clara, CA, 2000, pp. 164-173.

3. T. B. Jones, *IEEE Engineering in Medicine and Biology Magazine*, 2003, **22**, 33-42.
4. H. Morgan and N. G. Green, *Journal of Electrostatics*, 1997, **42**, 279-293.
5. J.-S. Kwon, J.-S. Maeng, M.-S. Chun and S. Song, *Microfluidics and Nanofluidics*, 2008, **5**, 23-31.
6. K. P. Chen, J. R. Pacheco, M. A. Hayes and S. J. R. Staton, *Electrophoresis*, 2009, **30**, 1441-1448.
7. J. L. Baylon-Cardiel, B. H. Lapizco-Encinas, C. Reyes-Betanzo, A. V. Chávez-Santoscoy and S. O. Martínez Chapa, *Lab on a Chip*, 2009, **9**, 2896-2901.
8. C. F. Chou and F. Zenhausern, *IEEE Engineering in Medicine and Biology Magazine*, 2003, **22**, 62-67.
9. G. B. Zhou, M. Imamura, J. Suehiro and M. Hara, in *37th Annual Meeting of the IEEE Industry Applications Society*, Pittsburgh, Pennsylvania, 2002, pp. 1404-1411.
10. J. Suehiro, G. B. Zhou, M. Imamura and M. Hara, *IEEE Trans. Ind. Appl.*, 2003, **39**, 1514-1521.
11. Y.-K. Cho, S. Kim, K. Lee, C. Park, J.-G. Lee and C. Ko, *Electrophoresis*, 2009, **30**, 3153-3159.
12. B. H. Lapizco-Encinas, B. A. Simmons, E. B. Cummings and Y. Fintschenko, *Analitycal Chemistry*, 2004, **76**, 1571-1579.
13. B. H. Lapizco-Encinas, B. A. Simmons, E. B. Cummings and Y. Fintschenko, *Electrophoresis*, 2004, **25**, 1695-1704.
14. B. H. Lapizco-Encinas, R. Davalos, B. A. Simmons, E. B. Cummings and Y. Fintschenko, *Journal Microbiological Methods*, 2005, **62**, 317-326.

(535f) Microstructure and Pattern Formation in Suspensions of Polarizable Spheres in Electric Fields

Jae Sung Park and David Saintillan, Department of Mechanical Science and Engineering, University of Illinois at Urbana-Champaign, Urbana, IL

We use theory and numerical simulations to investigate the microstructure in suspensions of spherical particles in a viscous electrolyte under a uniform electric field. Dipolar interactions between particles arise from dielectrophoresis (DEP) as a result of the Maxwell stress in the fluid. The dynamics under these interactions are analyzed in the thin Debye layer limit, and the formation of chain-like structures is predicted in the field direction in agreement with previous studies. Interactions between finite-length chains are also analyzed, and are found to be attractive in the near field but repulsive in the far field. Simulations in large-scale suspensions in a thin gap are also performed with periodic boundary conditions in the directions perpendicular to the electric field. Results confirm the rapid formation of finite chains, which is then followed by a slow coarsening process in which chains coalesce into sheets, which eventually rearrange to form mesoscale cellular structures. If the particles are polarizable, they can also acquire an additional non-uniform surface charge, which can result in interactions by induced-charge electrophoresis (ICEP). When both DEP and ICEP take place concurrently, a phenomenon coined dipolophoresis (DIP), ICEP interactions are found to dominate and suppress chain formation. DIP interactions are instead predicted to result in transient particle pairings, which lead to the formation of short-lived clusters surrounded by clarified regions. The effects of DIP on suspension dynamics are also investigated using large-scale simulations with periodic boundary conditions, where the resulting microstructure is characterized using pair distribution functions and particle occupancy statistics.

See more of this Session: Electrokinetic Behavior of Microparticles and Nanoparticles: Fundamentals and Applications
See more of this Group/Topical: Topical 3: 2010 Annual Meeting of the American Electrophoresis Society (AES)

(535g) Dielectrophoresis at Conductive Liquid Interfaces

Zachary R. Gagnon, Department of Cell Biology, Johns Hopkins University School of Medicine, Baltimore, MD

Dielectrophoresis (DEP) is a popular technique for cell and particle manipulation at the microscale, and is commonly used to describe the motion of suspending particles or cells under the application of a non-uniform electric field.

The DEP force results from electric field induced charge accumulation at the cell or particle surface due to a discontinuity in the electrical conductivity and dielectric constant at the particle-liquid interface. During exposure to an AC electric field, the interface charges much like a capacitor – governed by conductive charging at low (typically < 500 kHz) frequency and dielectric polarization at high frequency.

Commonly, small submicron particles and the interior of most cell types have a higher conductivity and lower dielectric constant than their surrounding media. Hence, the resulting direction of the induced particle dipole often reverses when the applied AC field frequency exceeds the inverse relaxation (RC) timescale of the media-particle interface, known as the DEP crossover frequency (cof). Interfacial polarization, hence, plays an important role in dictating the magnitude and direction of the field induced DEP force exerted on a particle or cell.

Typically, DEP has been applied to particles, cells and liquid droplets in suspension. At its most basic element, however, DEP is simply a force that arises from unbalanced charging at an electrical interface. Here, I explore a different electrical interface - that of two or more adjacent liquid streams flowing within the confines of a microfluidic channel network.

Due to the laminar flow profile and slow diffusion timescale associated with microchannel fluid flow, two or more liquid streams can flow side by side without mixing. As both the conductivity and dielectric constant of aqueous solution can be readily adjusted with soluble salts and zwitterions, two solutions of different electrical properties can be allowed to flow side by side to generate an electrical liquid-liquid interface.

Much like a suspending particle surface, a microfluidic generated liquid-liquid electrical interface can also be polarized with embedded microelectrodes. Here, I characterize the unique phenomena that arise at polarized electrical liquid-liquid interfaces. Furthermore, I apply them to electrokinetic cellular manipulation applications and biological studies and demonstrate the potential for this new and flexible microfluidic platform.

AUTHOR INDEX

AUTHOR INDEX

American Institute of Chemical Engineers
3 Park Avenue
New York, New York 10016-5991

CURRAN ASSOCIATES INC
proceedings.com